IT'S A
FARMING
THING

Kaleb Cooper

QUERCUS

First published in Great Britain in 2024 by
Quercus Editions Ltd
Carmelite House
50 Victoria Embankment
London EC4Y 0DZ
An Hachette UK company

QUERCUS

A CIP catalogue record for this book is available
from the British Library

HB ISBN: 978 1 52943 788 1
Ebook ISBN: 978 1 52943 790 4

10 9 8 7 6 5 4 3 2 1

Designed and typeset by Julyan Bayes at Us-Now Design & Art Direction
Illustrations by Julyan Bayes
Printed and bound in Italy by LEGO S.p.A.

Papers used by Quercus are from well-managed forests and other responsible sources.

Gramp,
I hope I've made you proud
in everything I do

CONTENTS

Hello again!

I'm buzzing to bring you my new book. Back when I was in-and-out of school as a kid (but mostly out, because I was building up my farming business from scratch, which you can read all about in just a bit), if somebody had come up to me and told me that I'd be a published author in the bestseller charts, I'd have thought the Cotswolds had suddenly got a crack problem. But here I am, on the printed page. Or maybe, if you're a futuristic young gun, like me, an electronic one. I'm bringing you everything I know about the farming life, and why I'm so devoted to it.

It's funny, being a book person. People – and by 'people', I mean hobby farmers, because no one else I know would be daft enough – sometimes ask me if I've read this or that book – usually it's a book that's got something to do with farming.

'No', I always tell them. 'I don't read books. I write them.'

The reason I write books is because I love farming, and I love the countryside, and I want there to be some books out there written by somebody who knows those things from the bottom-up rather than the top-down. I don't have time to read books about farming – or anything else for that matter –

because I'm too busy, y'know, farming – which, I like to think, has taught me more about the subject than any amount of books ever could.

That's why I've called this one *It's a Farming Thing*. It's not because you, the reader, wouldn't understand. Just the opposite. I really hope you will, and I've done my very best to make that happen. If you still don't, that's on me, not you. But the countryside has a lot of people in it nowadays who really don't understand (no names mentioned, ahem).

And the trouble is, it's people like that who end up writing most of the books about farming, as far as I can see. They treat farming as if it's all a bit of an adventure or an experiment, and when it goes wrong they can go back to their nicely paid day job. Not me: this is my day job, and often my evening job, and my night job.

And that's OK, because this is the thing I love doing most in the world. Well, fine, second most. To be strictly honest, the thing I love most is getting paid for it.

The only thing that comes close to my love for farming, is getting to tell other people about it. I made my farming career happen for myself, and I'm really proud of that. But it was pure luck that I happened to become a celebrity, with people interested in what I have to say. My friends still find that incredible. 'We don't even care what you have to say, Kaleb,' they remind me. 'And we're your mates. Now there's all these strangers who want to listen to you. It must be the end times or something.'

So, I'd be stupid not to take the opportunity to tell everybody why I love farming so much any chance I get. And that's what this book is for: to look at every part of my life as a farmer and explain to people why it matters to me, and why farming matters so much to everybody. I heard an old American folk song the other day, which is all about how farmers work the whole year around, living on credit and getting ripped off by lenders and middlemen. Then the chorus says, 'the farmer is the man who feeds them all.' That's from a hundred years ago. Some things never change.

Other things in farming never change, and that's a good thing. Those are some of the things I talk about in this book: the character of farming people, the commitment to a way of life, the way you have to be truly dedicated to the work. It's not a nine-to-five job. It's not something you do just to pay the bills. Every proper farmer knows that, and I hope that if farming people read the book, they'll relate. As for non-farming people, I couldn't be happier if you choose to read this book. I hope you'll come to share at least a bit of my enthusiasm, and understand how pleased I am to be part of a tradition that literally keeps this country going.

There's so much to cover. I've started with food, because from beginning to end, that's what farming's all about: making sure everybody's got what they need on their plate. I also wanted to write about work, families, kids, animals, machinery, clothes, and more besides. All the things that make up a farmer's life. I've even dipped into art, because I do believe there's an art to farming, and I've taken a look at some of the wackier ideas you find in country communities. Especially since city folk started moving here.

I've even spent a while looking back at my childhood and tried to explain how I've got here. Although, I admit, it baffles me sometimes. But I don't worry about the past because there's always the next day, and always something else to do, and that's what keeps it fresh. So, yeah, it's a farming thing. And I couldn't be happier to tell you all about it.

Chapter One

Food

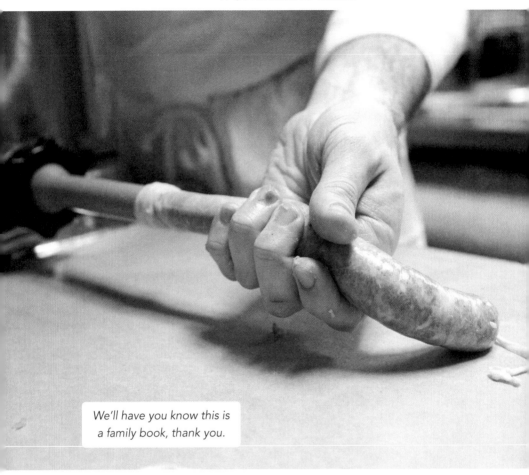

We'll have you know this is
a family book, thank you.

Food. Food, food, food. Food is the most important thing for
farmers. I suppose it's the most important thing for everybody.
Without it, we'd die, so that's about as important as it gets. But
for farmers in particular. We spend our lives producing it and
selling it. Then we spend whatever time is left over cooking
it and eating it. Nobody has a connection to food quite like
farmers do. For most people, suggesting that they 'know how
the sausage is made' is just a saying. For us, it's the literal truth.

Farmers know how everything is produced. Especially livestock: we've fed it and watched it grow, and at the end of it there's an animal that you can actually eat. When you know exactly what's gone into the animal and how it's been treated, it tastes ten times better. It's very rewarding. I'm not going to get into an argument with people who don't eat meat, but for those who do, you want to know the animal's been well treated. No one likes to kill an animal – not even when it's a horrible pig that you can't get into the pen. I've got one at the moment. I cannot physically get it anywhere, it's a complete menace, 250 pounds of bacon-flavoured attitude – and I'm not going to lie, when you push one of those off into the abattoir, it tastes even more amazing.

Trust us, it tastes a whole lot nicer than it looks.

I've got a veg patch full of runner beans, potatoes, onions. It's amazing to see that stuff grow all year round and to eat your own produce. People always tell me I'm full of beans.

Well, they should see my freezer. More beans than we'll ever need. It's unbelievable. It's not just farm-to-table, it's farm-to-self.

Like most people, I have three meals a day (not three an hour, as my mates sometimes claim), but I have them at different times from the rest of you and call them different things. Farmers live by a schedule that hasn't really changed much in hundreds of years. Everybody thinks farming's very advanced now, and yes, we've got modern technology, tractors with GPS and all of that. But the basics, and the timetable, have stayed the same. So, I'm really an eighteenth-century person on a twenty-first-century farm.

Kaleb's alarm clock.

Kaleb's snooze facility.

BREAKFAST

I have breakfast first thing. And I mean first thing – five or six o'clock in the morning. Unlike, ahem, some hobby farmers I could mention who think the working day starts at 10am after a full English, three lattes and the crossword. But it all depends on what kind of farming you're doing. When I was dairy farming, I might just have had a cuppa at 5am, worked until 10am, then eaten after that, once I'd built up a bit of an appetite. While I am a morning person – I have to be, I'm a farmer after all – my stomach's not a morning stomach. I like cereals in the summer months, but I've got a fascination with crumpets at the moment. With butter, of course. And jam. Obviously. And Cheddar cheese, which people might think is weird, but it all goes brilliantly together. People in Europe like quince jelly with cheese, apparently – I think they've totally got the right idea. I'll just keep eating crumpets with butter, jam and cheese until one day, suddenly, I can't face another crumpet, and then it'll be back to eggs – dippy or scrambled.

'Now, Mr Cooper, I see that you are with us today due to crumpet poisoning.'

DINNER

This is what most people probably call lunch. I have dinner at lunchtime. If I had it at dinnertime, it would be called supper. I used to always have a hot dinner (at lunchtime). Homemade soup – leek and potato or something like that – is amazing to warm you up when you've just come in from the fields and you have to go back out again straight after. Nowadays, because I run my own business, I just have something quick and easy because my boss (me) is a real hard case about his employees (me) taking too long over dinner (at lunchtime). So, it'll just be a toastie or a sandwich, using nice fresh bread, with Cheddar cheese or ham. A very basic version of a ploughman's lunch, which really ought to be called a ploughman's dinner (at lunchtime), but I don't write the pub menus.

Chopping board? Very fancy! Next you'll be serving food on a roof slate!

TEA/SUPPER

This all depends on what time I finish the working day. In the harvest season, I normally have supper because I eat quite late, sometimes not until after nine. I'll have done so much manual labour that I'm starving. I'm so tired I could go straight to bed, but

I can't sleep if I don't eat. My stomach won't have it. It's an evening stomach, and if I don't feed it, it nags me all night – it literally grumbles at me. In the winter, I have tea, which is at five or six o'clock in the evening. And I seem to put on a lot of weight. I think it is probably in my DNA. A bit like those people in the Arctic who kill a whale when the long winter approaches and eat as much of the fat as possible to bulk them up. Or I may be part bear – I certainly get an urge to hibernate, although when you're a farmer, fat chance, so to speak.

For tea or supper, it's usually a stew (I love me a stew done in a slow cooker) or a roast. Just give me a plain roast beef dinner. Lots of gravy. Horseradish. That's what it's all about. Horseradish, pepper and mustard are about as strong as I can take, flavour-wise, although I do love a bit of chopped raw onion on a steak. I can't handle spices, which is a pity as I'd love a good curry from time to time, but I'm just not built for it. If I risk it, I have to put a toilet roll in the fridge the night before for the morning after.

Sometimes, I think I'm part bear, bulking up in the winter to carry me through the long nights.

HOME COOKING

I'm bloody awful at cooking. My other half has banned me from the kitchen. I know some men have that thing where they do stuff badly on purpose so the woman in their life has to do it instead, but I swear, this isn't that. The other day, I reheated some soup, then I went off to sort something or other out and forgot about it. It was a disaster. I scorched the pan and ruined the hob. I'd love to be good at cooking. I watch all the cooking shows, *Masterchef* and the rest, and videos on YouTube – it fascinates me – but God help me, I physically can't do it. So, I just stick to growing stuff. In the summer, we'll have salad leaves from the garden. Then there's runner beans, broccoli, peas and potatoes. We grow so much fruit and veg, our kitchen looks like it's having its very own harvest festival for a few months every year. It's wonderful growing your own food – I just can't be trusted to cook any of it.

'Kaleb, have you been making beans on toast again?'

THE KALEB COOKBOOK

Since I've been on telly, I often get people calling me up and asking me for a recipe for a cookbook or an article – but it's a bit like asking the Incredible Hulk for advice on tailoring. I only know one recipe and it's for radishes, which are my favourite snack when I'm working.

Radishes à la Kaleb

* Take one radish.
* Bite the top off.
* Spit it out the back window of the tractor.
* Eat the radish.
* Repeat.

Haute cuisine ahoy.

AND TO DRINK, SIR?

I normally have tea. I don't drink much coffee anymore because I overdid it when I was younger, milking cows bang early. I started getting evil headaches – it got so bad I went to the doctor's and had loads of scans, only to find out I was addicted to coffee. Coming off it was horrendous. I had to go pure cold turkey – and not the good kind you put in a sandwich for your dinner (at lunchtime). I felt how I imagine Keith Richards must have felt, or the guy out of *Trainspotting*. OK, I didn't see terrible things crawling on the ceiling, but that's because I was too busy climbing up the walls myself.

Don't get me started on nut 'milks'. You can't milk a nut.

'Cider and combine harvesters? You're speaking our language, Kaleb!'

So, now I just have a cup of tea in the morning. Or apple juice in warmer weather. Sometimes a mix of apple and orange juice, half and half, which sounds strange, but it's amazing.

I might have a can of Coke with dinner (at lunchtime). At tea or supper (at dinnertime), maybe something stronger but I'm not much of a wine drinker. To quote The Wurzels, and there's never a wrong time to quote The Wurzels, I am a cider drinker. I'm very much a cider drinker. I am so much a cider drinker that I'm having to be careful to be a lot less of a cider drinker. I know it was normal in the old days for people to drink beer or cider at any time of the day, because it was safer than the water – or at least, that was their story. It just sends me to sleep, though. The kind where you wake up and you don't know what time, or day, or century it is, or who or where you are. One thing I'll admit, though, I do like a nice, slow drink of cold cider when I'm corn-carting (which is pretty much what it sounds like, only these days instead of a horse and cart, you're on the combine harvester). It's more of a social thing, corn-carting, a get-together, when there's a few of you on the combine or the tractors, trundling up and down, and you raise your drink when you pass each other – 'Cheers, mate.' Obviously, you don't want to be drunk in charge of a combine harvester, or very bad things can happen, but a little of what you fancy helps get the grain in.

CHRISTMAS

We've only got one proper feast day in our country, which is, of course, Christmas. Maybe that's why we make such a big deal of it – and spend what feels like a quarter of the year preparing for it. I love Christmas. It's my favourite time of the year. Everyone wakes up in a different mood, a mood of joyfulness, and the whole family teams up to look after all the animals – because obviously animals don't take the day off, so farmers can't completely either. I don't have breakfast on Christmas Day. At least, not strictly speaking, because there's so much food being prepared in the kitchen, I just pick at it.

When it comes to serving up time, people often wonder why the turkey's got no wings on it. No use telling them it's one of those flightless turkeys, either, because they just point out that turkeys can't fly – so I try to change the subject instead. We go all out on Christmas dinner (at lunchtime – see, it's not just me) – loads of meat and all the trimmings, which we've grown ourselves. We have a gammon, a joint of beef, a flightless turkey and a goose – which I'm telling you is the king of Christmas meats, and which also gets what's coming to it. I've never forgiven the goose that attacked my, um, delicate parts and almost ended my family line. They say revenge is a dish best served cold, but anyone who's tried potatoes roasted in goose fat can tell you the opposite is true.

Let's see you honk and peck your way out of that one.

I don't have much of a sweet tooth, so Christmas is also the only time of year I get into desserts. Banoffee pie, especially. Well, you've got to get some sort of healthy fibre into your meal, haven't you?

*Banoffee pie –
technically speaking,
one of your five-a-day.*

FEASTS AROUND THE WORLD

I really like the idea of Thanksgiving. I actually think every country should have one, not just the USA – to say thank you for all the food we produce to a very high standard. We need to appreciate how lucky we are. Plus, I can rear some more turkeys and make some money off it.

I also like the midsummer festival they have in Nordic countries. I'd love to celebrate midsummer, but it's peak season on the farm and I haven't got time to blink let alone take part in a festival. The food sounds all right, too. Well, mostly. Herring, salmon, potatoes, that's all good, but fermented shark? In Iceland, they bury it for three months, because that makes it less toxic. Anything that gets less poisonous after three months in a hole is probably not a great idea in the first place. But I appreciate that Iceland wasn't the easiest place to find food back when they came up with that idea. Otherwise, why would you do that? Why?

'I'm not too keen on the whole idea myself.'

But the place I'm really in tune with is Greece. Lots of countries around the world traditionally have a big nosebag in the lead-up to Easter, just before Lent. That's why we have Pancake Day, eating up all the foods we'll be denied until Easter. In Greece, though, they get serious about it. They have one whole week that they dedicate to eating meat, followed by another whole week dedicated to eating cheese. When I heard about that, I

thought, I love Greece, it's going to be first place I go when I get a passport. I mean, yes, it's the birthplace of democracy, science and philosophy, and that's all well and good – but a national cheese week? That's what I call civilization.

What a friend we have in cheeses.

Chapter Two

Pets

If you ask me which animal makes the best pet, I'm one hundred per cent going to say a dog. I'm such a dog person it's unreal. I hate cats. Cats are just there to piss you off. They do it in ways a dog would never think of. Yes, a dog may not listen to commands sometimes, but at the end of the day it's very sorry for not doing what you asked. But a cat. A cat knocks things off shelves, and doesn't give a damn. I know some people say this shows that cats are a lot smarter than dogs, and a lot cooler, but I'd suggest that those people are just a lot more gullible than dog owners. A cat behaves with complete contempt towards its owner and yet they worship it as if they were ancient Egyptians. All the while the cat's thinking, 'Treat 'em mean, keep 'em keen.'

Man's Best Friend.

'Revere me, human mortal. Also, my food bowl has been half-empty for at least seven minutes.'

Can a cat herd up to a hundred sheep? Or push a lot of cows through a gate you're struggling to get them through? I suppose what it comes down to is that dogs do what you want, and cats do what they want. Which is all very well for cats, but then, why would you want to own one?

Attempts to film the series One Man and His Cat *were quickly abandoned.*

WORK AND PLAY

There's a difference between pets and working animals, although, I say, get yourself a dog that can do both. Like my Australian collie, Jess – she's in retirement now, but when she was working, she really worked. She was like me actually – I'm twenty-five years old, I go out and work my a*** off, but I come home and I'm a family man. The best dogs are the same: up in the morning, working all day, but soon as you're back home they can rest and be the family pet.

'I'd love to play fetch with you, but I really do have to sort out these accounts.'

When I was a kid I had a dog called Diesel. He was the perfect example of a dog who could do both – the best ratting dog I ever had in my life. He could kill vermin all day long, then come home and be the most cuddly dog there ever was. I remember my brother getting a hamster and it was really loud at night, always clattering in its wheel. Until one day the hamster went … missing. My brother was distraught. I had to pretend it escaped. I couldn't tell him what really happened. I didn't blame Diesel. He did his best to respect a fellow animal in the house, even if it was a rodent, but I think the night-time noise got to him in the end, and he went, 'Sod this', and that was the end of that.

TOWN DOG, COUNTRY DOG

Just as there are two different types of people on this planet, so there are two different types of dog. My boy and a city boy would be different. My boy can eat a bit of dirt, walk off where he wants and climb an old tractor. But a city boy ... well, not likely. In the same way, if there were two dogs in the same room, one from London and one from the countryside, I can guarantee I could tell you which is which. And not just because the London dog is wearing white trainers.

'Don't be ridiculous. I only wear dark-coloured trainers to the countryside now.'

There are two different types of owners, as well. There's the person that walks the dog, like me, and there are people that get walked by the dog. You see them getting pulled along with their arm stretched out and zero control over where they're going. The dog is completely in charge. My dog will never be on a lead, and whenever you call him back, he's right there. I saw someone I know the other week and he was speaking to his dog in Welsh and the dog did what he told it to. Or I assume it did. The first thing I thought was, how on earth does the dog know Welsh? The second thing I thought was, OK, but then how does my dog know English? Maybe I should try out some other languages on him. 'Bonjour!' Then I started to wonder if dogs have accents, too. I bet my dog has the same accent as me, but only other dogs can tell. My mind is a crazy place.

'Et bonjour a toi, mon meilleur ami!'

WHY THE LONG FACE?

Another working animal I have a lot of respect for is the horse.
I'm not saying I *like* horses. But I like having them around,
even if they are total hay burners – it's like shovelling coal
into a furnace, it never stops. I especially like old-fashioned
shire horses and the Suffolks – after all, they're the original
horsepower. They changed agriculture and made it possible
to produce food and get it out to people. Then, obviously,
they got replaced by the tractor, which is the best thing that
ever happened.

Horses. The farming OG.

OF MICE AND MEN

I do have to give some credit to cats here, because when it comes to the history of farming, they're all about the pest control. If rats and mice are eating the grain you've stored, then you're going to need a cat. Let's face it, cats just love killing things. They're nature's perfect killing machine – and it's a good thing we're too big for them to take down, and too useful for their own sinister cat agenda, or we'd be cat food, no question. So, there's an uneasy alliance there. As long as they've got small, squeaky things to hunt and eat, they'll be happy enough. This is why I'm wary of going anywhere that's got big cats around: to them, we're the small, squeaky things. You do hear stories and see supposed photographs of panthers and things like that roaming the British countryside – but I reckon that's mostly people with lively imaginations who've been

on the scrumpy and spotted the next farm's moggy out for a wander. And, let's face it, those photos always make the pictures of Bigfoot or the Loch Ness Monster look super clear and believable.

What if a sighting of the Big Cat was a just a … you know … a cat?

WEIRD PETS

There are people whose whole personality seems to be centred around keeping weird animals as pets. My mum went through this phase. I thought it might be something to do with the menopause, although I'll probably get into trouble for saying that (and I think I'm out of my depth) ... She had a tarantula once, when I was fourteen or fifteen, and I remember one morning she came in and said, 'Kaleb, the spider's out.' I wasn't fully awake yet, and I thought I must be dreaming. Then I got up to go to work and said to mum, 'I had a funny dream last night, that you told me the tarantula had escaped.' And she said, 'No, Kaleb, it has – I can't find it anywhere.' *What?!* It was a Chilean rose tarantula, which can shoot fur into your eyes and blind you. Why on earth you would have that in your house, I do not know. But her house, her rules, her insane pets – you've got to respect these things. Eventually it was found under the sofa. Although not by me, luckily, because that was one of the first places I thought of hiding.

After that, mum decided to go for something a bit safer and more normal. So, she got a python. When blokes have a midlife crisis they get a sports car or a ponytail, but for mum, it

The good news is that they only shoot their blinding fur when they feel threatened. Or irked. Or bored. Or just for the hell of it.

was a giant predatory snake that could suffocate you in your sleep. She'll kill me for saying that … but then I've already survived a poisonous fur-flinging tarantula and a man-eating reptile with coils of death, so I fancy my chances. She never used to tell me she'd got these pets, either. I'd get in from work and they'd just be there.

Our house was very busy because, when I was thirteen, my mum bought me three chickens and I started my first business selling eggs to locals. Before too long, I had 400 chickens. I bought some incubators and was hatching chicks to grow the business. I remember taking a fresh brood upstairs to see my mum. She was having a rest in bed and there was a python lying on top of the duvet. I said, 'Why have you bought that? I don't have that many chicks that die, Mum, so what are you going to feed that thing?' I never got close to the python – and I don't mean emotionally. I wouldn't go anywhere near it! One of the good things about our country is that, usually – *usually*, Mum – you don't have to worry about animals that want to choke you, blind you, poison you, eat you or any combination of the above, and I'd like to keep it that way.

'Why, no, I definitely haven't seen Kaleb anywhere – why do you ask?'

39

It could have been worse – at least my mum never had a grizzly bear, although that was probably only because she couldn't get hold of one. There's a guy in Montana who rescued a grizzly bear as a cub, and now it weighs 900 pounds and follows him everywhere. He says that like it's a good thing, but really, is it? Is that loyalty, or just stalking? Maybe the bear's thinking, 'One day, you're gonna drop your guard, man, and then ...'

'Any moment now . . .'

Then there are folk who have crocodiles and alligators as pets. Have these people never watched a single *Jurassic Park* movie? Those things are basically dinosaurs, and nothing good ever comes of hanging around with them. OK, strictly speaking, those guys and dinosaurs both come from a group called the archosaurs, which apparently means 'ruling reptiles'. But there's a big clue there: you might *think* you're at the top of the food chain, but they *know* they are. They were around for tens of millions of years before humans, and they haven't survived this long by being anybody's pet. Any self-respecting alligator or crocodile will be thinking, 'I just want to go and swim in a natural lake and get away from your bullsh**. And if I have to eat you for that to happen, well, that's a sacrifice I'm willing to make.'

'I'm so sorry it – *burrrp*; excuse me – had to come to this.'

PET FOOD

Crocodiles and alligators aside, if we're honest, it's more likely that we'll end up eating our pets than the other way around. I don't mean our own pets. But one person's pet can often be another person's dinner, especially in different cultures. Sometimes even in the same culture. I've heard of a French poet who used to take a lobster for a walk on a ribbon – which sounds totally like something a French poet would do. 'Why should a lobster be any more ridiculous than a dog?' he said. Well, I can think of a few reasons, at least one of which involves garlic butter. But that's the point: it was his pet and he never would have eaten it. Or look at rabbits. I know they're still eaten, but they used to be a staple – there's even that famous song about it, 'Run, Rabbit, Run', which was really popular in the Second World War. It's got a line asking how the farmer will get by without his rabbit pie. Well, I mean, I certainly can – rabbit's not a favourite of mine, either as a pet or a meal – but if I'd been around in the war, with rationing, I bet I'd have been thinking twice about that. In South America, they eat guinea pigs, which isn't all that different. And in Cambodia, they eat deep-fried tarantulas, which might upset my mum, but I'd say it's getting your retaliation in first. Again, just like Britain in the war, you have to take your protein where you can find it, so fair play.

You couldn't... Could you?

42

THE PERFECT PET

The top three qualities to look for in a pet are obedience, loyalty and trustworthiness. For example, we've got a pet cow called Heidi. She's a Highland cow. And she's the complete opposite to all those three things. When you get into the pen, she'll come over for a two-minute scratch, love you for those two minutes, then go, 'You know what, I've had enough now', and just round on you and kick you. I like her, though. She's got an attitude on her. Then, back when Taya and I got together at sixteen, I bought her a pet lamb for a tenner. I thought, she'll be a happy girlfriend, she's got a pet lamb. She can rear it, she can keep it in the conservatory at the house. I bought a girl a lamb so I could breed from it and earn some money. It was all good until I got too busy and I fell out of love with sheep. So, now we've got eight pet sheep. Eight. It kept breeding, and every single lamb it had was a girl. Of course, Taya named each one and after that there was no getting rid of them. They make Heidi look as obedient, loyal and trustworthy as a border collie. So, I brought this on myself.

Sammy the lamb: the way to a woman's heart.

Chapter Three

Family Traditions

There was a Russian writer called Count Tolstoy, who I like – firstly, because he sounds like a really cool jazz pianist (I don't even like jazz, or pianos, but fair play, it still sounds cool), and secondly, because he was very into agriculture, and that's enough for me.

He was posh as anything, but in his later years he took up the life of a rural farmer and … hang on, that sounds a lot like Jeremy. I wonder if Tolstoy also had to get a proper farmer to come and literally dig him out of a hole every five minutes. Anyway, the point is, Tolstoy said, 'All happy families are alike; each unhappy family is unhappy in its own way.' Sorry, Leo, but I'm not sure you know what you're talking about there.

'What do you mean, mini me?'

My family is pretty happy – but if every other happy family was like ours, it would be pure chaos. You only have to look at our family traditions to see that.

46

Kaleb (right) and brother, successfully not destroying anything. Yet.

WE WISH YOU A MERRY CHRISTMAAAAARGH

Some of our family traditions seem like pretty normal traditions. Christmas, for instance – what could be more normal than that? Yet, in my family, for some unknown reason, a few days ahead of the big day itself, everybody gets together for a meal in a pub. It's like we're having a rehearsal dinner for Christmas, and all that happens is that everyone has a lot to

'Well, I didn't invite him, and you didn't invite him, and neither of us knows who he is – but he's certainly getting into the spirit of things.'

drink and ends up getting into arguments. Except for me – I'm a lover, not a fighter.

Usually if I see an argument happening in front of me, I just walk away. But if I'm sat round the pub table I can't do that, of course, so I have to sit back and watch it, thinking, 'You're all mad.' Then we all get back together on Christmas Day itself, and we have to feed everyone and argue all over again, and I have to sit there and watch it and think, 'You're all still just as mad as you were two days ago.' Don't get me wrong, I love my family, but I'm sure they could manage with having a mahoosive ding-dong just once a year instead of twice in the same week.

GO KALEB, IT'S YOUR BIRTHDAY

We normally push the boat right out on birthdays and make a big deal of it – presents, cake and a pint of cider if the weather allows. I've always struggled a bit with birthday cards, though. When I was a little kid, if I got a card, I opened it to see if there was any money in it, then put it down and never looked at it again. So now, when I get a card for someone, I can never think of anything to write. I always want it to be something different or interesting, but there never is anything different or interesting to write, is there? It's got to be, 'Dear [whoever], Happy Birthday, All the best, Kaleb', and I think it must look pants. My mum and my nan love a card, though, and dad likes a card as well, so I'll be in trouble if I don't do it. I suppose I could always try putting some money in ... but that's crazy talk.

'It's just what I wanted!'

SO LONG AT THE FAIR

When I was growing up, we'd go to an agricultural show every year – and I'd look forward to it for weeks. It's like a showcase for everything farming related, all in one place – the stalls, the produce, the animals, the equipment and, of course, the actual farmers. If, like me, you were obsessed with farming, it was the best place in the world. We used to go to the Moreton-in-Marsh show, or the Blakesley show, which, given it's almost an hour away, proves it was a big deal – for me, that's like an epic quest from *The Lord of the Rings* or something. Even the Moreton one was pushing it a bit, and that's almost round the corner.

I'm not sure if this is business or pleasure.

It's funny, but if you go to an agricultural show for fun, you end up doing mostly business, and if you go for business, you

Where's my brother when you need him?

actually have more fun. Basically, if you go for leisure, you'll find you spend the majority of the time talking business with people. I know everybody, because I'm an agricultural contractor, and they all want to talk shop. Taya gets annoyed and just walks off now. But if you have a stand there, to promote your business, you say, 'I'm just going to go and get some food', and then spend most of the time walking around seeing everything else, rather than stay on your own stand.

I still think those agricultural shows I went to as a youngster are a big part of the reason I'm a farmer today. This is one of the family traditions I'm excited to pass on to my own family. When I was a kid, the family would all go together – or whoever could be free that day – and I'm not from a farming family. Often my mum would be working, so my nana and Gramp would take me and my younger brother. My brother loves anything mechanical. He can take anything apart and put it back together again. I mean, so can I, strictly speaking. The difference is, when he does it, it still works afterwards.

So, my brother would spend all day looking at the old machines, while I'd be looking at animals, especially the chickens, which were my first business, and the bantams, which are still my favourites. If my mum did manage to come along, she'd be looking at all the crafts, which she loves. So, there was always something for everyone. I think the highlight for me was when my nan and I went to Malvern, for the Three Counties show. It was amazing. There's only one person in the entire world who likes eating cheese more than I do, and that's my nan. Or maybe it's the other way around. I would gladly take part in a study to find out, so if anyone wants to set one up, we'll both be there like a shot – purely in the interests of science, naturally. They've got a cheese room at Malvern, and she and I would just go up and down it all day, eating free samples of cheese. 'Have you tried this one yet?' 'No – and what about that one over there?' How we ever managed to walk out of there, rather than being wheeled out, I'll never know. A visit to that cheese room is a family tradition I plan to keep up until the day I die, preferably from eating too much cheese.

Mmmm. Cheeeeese.

KIDS TODAY, EH?

Community events are really important, I think. Like the Chipping Norton festival in summer and the Christmas lights switch-on in winter, when the mums and dads make all the youngsters go out. Normally, they'd be sitting indoors on their Xbox or scrolling their phones but their parents force them out of the house, and they see not only all their mates from school but other people from the local community too. And it feels safe. I was once asked to go and turn the Christmas lights on and say a few words on stage in Chipping Norton. Nowadays, I do public speaking all the time and it's fine. But as soon as I got up on that stage I froze because I knew everyone in the audience. I couldn't spit a word out.

A lot of traditions have been killed off by health and safety, and to be quite honest, whatever some people say, I don't think that's always a bad thing. I don't know if people were just tougher back then, or if human life was cheaper, but if you read my last book you'll know how many cultural traditions and challenges involve things like throwing yourself down a hill after a wheel of cheese or jumping off a cliff with a homemade set of wings strapped to your back. So, anything that sends kids outside, blinking, into the sunshine, or to look at some pretty decorations on a winter's night, without ending up in hospital, gets a thumb's up from me. I suppose some of them might manage to electrocute themselves through a combination of cider and fairy lights, but then that could happen at home on your own, and I don't see anybody trying to make it part of our communal folklore.

Anything that gets them outside…

REMEMBER REMEMBER (EVEN IF YOU'D RATHER FORGET)

My dad is terrible with fireworks. Absolutely terrible. But that's never stopped him. He's always first to volunteer to organize a fireworks night – another family tradition of sorts – although I use the word 'organize' in its very loosest sense. There's a local youth club at Old Norton Hall nearby and one year my dad was chairman of the club, so he was in charge – and again, I use that term in an even looser sense than 'organize' before – of their whole fireworks display. Of course, my brother and I went along to support him, because it was great he was putting on a fireworks night for all the kids. Or rather, it was great in theory. He lit the first firework, turned around to walk away from it, and the firework fell over and went off at the same time. Luckily, it hit me in the leg – luckily, that is, because it didn't hit someone else. I think even my dad was relieved he hadn't managed to injure anybody else's child. 'Oh, cheers, Dad!' But at least he wasn't being dragged off by the authorities, or so I thought at the time. In hindsight, it might have been better for everybody – him included – if he had been, but I was only around ten so I didn't have that perspective. All I had was a scorched leg and a temporary (thank God) limp. My dad is now banned from handling a firework within a five-mile radius of me. I sometimes think the reason I don't mind getting knocked about by farm animals so much is that it's a doddle in comparison with what my family got up to purely by accident when I was growing up.

Ooooh! Aaaah! OWWWWWWW!!!!

Recently, I turned an old horse box into a cider trailer and me and my dad took it to the fireworks night at one of the farming shows. But when the fireworks started I panicked and just hid behind dad. So, he's ruined fireworks for me. He kept on giving me cider though, which helped a bit. OK, it helped a lot, but then cider helps everything. It was like my least and most favourite things in one place. On paper, cider and fireworks may not sound like an ideal combination, but nobody was hurt. For a change. Sometimes a break with tradition can be a good thing.

Cider: Making everything better since 3,000 BC.

NEW TRADITIONS

The thing about traditions is that they have to start somewhere, and now I've got kids, I want us to create some of our own. There's a place called Blenheim Palace not far from us and I like to take the kids on the Christmas light trail there. They light up the palace and the gardens with twinkly little lights and on the website it looks like a brochure for the most magical Christmas imaginable – but you try taking a two-and-a-half-year-old feral farm kid into a palace. It gives me nightmares. He's touching everything. He's pulling everything off the wall. He's trying to get to the chandeliers even though he's only tiny and they're way up on the ceiling. Despite that, we do it every year, the kids love it, and I hope we'll continue doing it for years to come. They do a Halloween lights trail too, but the boy's not afraid of anything, so that makes it a bit boring. For him, I mean. It's not boring for me because I'm afraid of him – specifically, of him destroying everything. And if that's not a family tradition, I don't know what is.

The Christmas lights at Blenheim Palace.

The Christmas lights at Blenheim Palace after Kaleb's family have been for a visit.

Chapter Four

The Weather

If you're British, then you'll know that the weather is the go-to topic for polite conversation. This is a problem for me though, because whenever anybody tries to talk to me about the weather, my immediate instinct is to shout 'F*** THE WEATHER.' I don't, of course, because they're only being friendly. It's not their fault that my life revolves around the bloody weather – and I hate it. What's small talk for everybody else is very, very big talk for me. I've got five weather apps on my phone, and I check them approximately 1,574 times a day. Most people spend something like four hours every day on their phone looking at social media – I spend that much time looking at weather apps, in the hope that one of them will tell me something I'd rather believe. Then the weather

Yes, but it's a lovely day on my phone.

just happens anyway, and it's usually rubbish, and all that time spent on the weather apps has been completely wasted. But I can't stop myself. You'd think that, if there was anyone who understood cause-and-effect, it would be a farmer – you do A, then B happens – but I still keep thinking that my phone will somehow make the weather better.

The ironic thing is that most farmers are probably way better at predicting the weather than the BBC or the Met Office or whoever. A lot of those old sayings – 'Red sky at night' and so on – are based on real knowledge. Not that red sky at night really does mean shepherd's delight, because the only thing that's ever going to delight a shepherd is changing their job and not having to work with sheep ever again.

Delightful, except for the sheep part.

But still. You know what things to look out for – low clouds, high clouds, wind speed, pressure. Everybody who works outdoors will have a stronger feeling for the weather than people who work indoors. But it's especially true for farmers, because it determines how much money I'm going to make, or if I make any at all. Your livelihood is decided by something everybody knows you cannot control. Not even if you're a mad scientist with a weird machine in a Kate Bush song. Somebody showed me that video once, and I thought it was just a made-up story, but it turns out he was a real person called Wilhelm Reich who built a contraption to make it rain. The song goes, 'I just know that something good is going to happen ...' It didn't, though. Instead, the feds came and took him away, which if you ask me is fair enough. Now, if he'd been trying to stop it raining they should have given him a medal.

At least having five weather apps on your phone is a bit more manageable than Wilhelm Reich's device.

THE FOUR SEASONS

In an ideal world, which is very much not the one I live in, this is what the best weather for farming would be across the seasons. In spring, I need it to be warm and damp. I would say 'moist' but I don't like the word. It sounds weird when I say it … 'moist'. Even writing it feels slightly creepy. Moist. Moist. Moist. I'm sorry I had to put us both through that. So, yes, warm and damp for the greenery to grow – it makes the heads of corn strong and heavy. This really *is* starting to sound like a dodgy novel now: *Fifty Shades of Corn* or something like that. Anyway, we sell corn by weight, so the heavier it is, the more money we get paid, which to be honest I find much more exciting than the idea of being tied up by some businessman.

In the summer months, we need it lovely and warm, and the sun to be shining, so we can make our hay. Again, what to other people is just a turn of phrase – making hay while the sun shines – for us, is our living. We want the long summer days to be clear and sunny to bake the grass for us, but even more important is the wind. The wind dries things out more than anything, so we need a nice warm wind blowing through the hay or corn. And the animals like that, too, as long as it doesn't get too hot and bake all the moisture out of the ground, because the grass is still growing.

Then, in the autumn and winter months, it can rain and snow as much as it wants. I actually prefer the frostier days. I love waking up on a frosty morning knowing I can drive a tractor across a field and not make a mess because it's crisp and hard.

An actual picture of Kaleb, quite literally in his element.

STYLING IT OUT

A farmer needs to be one hundred per cent weatherproof, which is particularly tricky for me because, as anyone who knows anything about me can tell you, I love doing my hair. I don't think many farmers care about their hair as much as I do. My hair is my hobby – and, while I've still got it, I'm going to make the most of it. But hair and weather do not get on. So, if I know it's going to be raining, I won't put that much hairspray in my hair.

'Heavy showers all morning – there goes my perfect ploughing lewk.'

If you want to know how to dress for the weather, especially for cold weather, ask a farmer. Hand-warmer packs are wicked. Your hands and your feet always get cold when you're out working. I don't care about the quality of my jeans, but good wellies or boots are crucial – even more important than a warm, waterproof coat. Never scrimp on your footwear. If you buy a cheap set of wellies and you have to spend twelve hours a day in them, you'll totally understand the meaning of a false economy. Also, you might think you should go for a very thick sole, or a heavily insulated welly, but that's not always the case. Your feet may sweat, and the sweat gets cold, and then your feet are covered in freezing moisture. You want a balance with something breathable. I always wear a thin pair of socks, and I never double them up. Not having the movement restricted in your feet allows them to stay warmer, so, whatever kind of boots you use, they should always have a bit of wiggle room.

Kaleb's new boots have been delivered.

BAD WEATHER

If you want to know what the worst kind of weather is for farmers, just take a look at 2023. It was, to put it mildly and without swearing, exceptionally tricky. We had a really sh** wet spring – I'm sorry, but it turns out that putting it mildly and not swearing simply aren't options when it comes to that utter b****** of a year – so nothing that we planted would grow in the ground. Then we had a really, really wet summer, so we couldn't get the combine in on the correct moisture levels – which is the amount of water in the soil – or do our field work, or bale hay, or bale straw for our cattle. For instance, harvesting wheat needs a moisture level of fourteen per cent, and we were at seventeen per cent, which means you'll have to pay a drying charge to somebody to get the wheat dry. And the rain was also driving the corn onto the floor, where the combine can't reach it. Then comes winter and it was still wet. By November, we'd already exceeded our average annual rainfall, so everything that followed was extra. I'd find myself sitting at the top of a hill in thick fog and drizzle, unable to see my hand in front of my face, let alone actually get anything done.

Basically, 2023 was a write-off. I reckon we probably lost thirty to forty per cent of our yield. I was so fed up with it, I was glad to see the back of it. But the lucky thing is, we get to go again, don't we? Every year, we get another year, to hope we can make it better.

Burn the whole year down. Not that you'd have any chance of setting light to it.

CLIMATE CHANGE

Don't worry, I'm not one of those people who claim this isn't happening, or it's being exaggerated. Obviously, there's scientific data which proves climate change is real, and just speaking for myself I have definitely seen an effect on the weather in my relatively short career as a farmer. We haven't had a good amount of snow in years, and in 2022 we had really hot weather continuously. But what really frustrates me about the whole thing is whenever people start talking about climate change, it's always the farmers who get the blame. The other day I saw a travel journalist being interviewed, and he was saying how everyone should stop eating beef and we shouldn't have cows on farms. Think about that for a minute: *a travel journalist*. Mate, I bet you've done more air miles than any cow has.

Next stop: over the moon.

There are so many different ways we can fight climate change, and I'm not saying farmers shouldn't play their part, but out of all the things involved, we should just remember, farming is the only one that we quite literally cannot live without. It doesn't matter if you've got three iPhones and an Apple watch, or what Xbox or PlayStation console you own, what car you're driving or what house you're living in: the day there's no food on your table is the day you need to worry about. There are a lot of people in the world and far too many of them go hungry, but we still manage to feed most of them – and for that, you can thank a farmer. And yes, that includes me.

EXTREME WEATHER

The most extreme weather I've ever experienced was 'The Beast from the East'. I was only seventeen and I owned a few sheep – and, of course, the only thing worse than owning a few sheep is owning a lot of sheep. It was lambing time and I was out in the fields – and I discovered that they were lambing under the snow. I remember putting one sheep over my shoulders and holding two lambs in my arms, trying to carry them somewhere warm and dry. I had to do it all on foot – I couldn't get a tractor through because the snowdrifts were so high. And if you can't get a tractor through, you know it's serious. Some of the drifts were ten- or fifteen-foot tall.

Er, Kaleb, I think I'm going to need your help…

All that said, I quite like it when it snows, because you can earn a fair bit of money pulling idiotic people's cars out of a snowdrift after they've gone out to buy a pint of milk. So, I suppose you can say it's swings and roundabouts.

I've also seen a meteor shower coming in, and I thought that was our Tyrannosaurus rex moment. One minute, top of the food chain, the next extinct. And getting hailed on in the middle of summer was pretty weird too.

Yeah, 500 metres is close enough to this, thanks.

Then there's lightning, which is an occupational hazard for farmers. I was muck-spreading on the tractor one evening, in pitch-black conditions, when lightning struck a gas-production plant nearby and the whole thing went up. I could see everything as clear as daylight. You think you're safe on a tractor, because the tyres insulate you from lightning …
But they don't insulate you from a huge exploding fireball of flaming gas, do they? I rang my mum and told her I loved her. I was about 500 metres away, but it was so terrifying that I really thought it was the end of me. Of course, if I tell the story in the pub, I say I was about four metres away and I could smell the singed hair on my eyebrows.

REALLY EXTREME WEATHER

Despite the snowdrifts and the exploding fireballs, when you think about it we get off pretty lightly farming in the UK. If you're farming in the Great Plains, where Americans grow a lot of their crops, tornadoes are so normal they actually have their own season and you have to take out insurance against them. And if you're near the coast, you get waterspouts, which are basically tornadoes made of water, and that sounds even worse. Yet another reason to never go anywhere near the sea. It makes me glad to be seventy miles inland. Then again, inland you can get fire devils, which are like fiery tornadoes – if there's anything that would make you nostalgic for a waterspout, it's got to be that. And that's before you look at things like raining fish and frogs, or frost quakes, or thundersnow. I mean, thundersnow? That sounds like some sort of sex movie. Apparently, thundersnow is so bad it even scares people in Scotland, and people in Scotland are never scared of anything, let alone the weather. They can have a mahoosive blizzard that buries the house and they're like, 'Och, it's a wee bit chilly, I might put on a cardie.' So, if I ever complain about anything that happens in the Cotswolds, just say 'thundersnow' to me and I'll shut up.

I mean, we're scared already.

WEATHER SAYINGS

Red sky at night, shepherd's delight …
red sky in the morning, shepherd's warning

All a red sky at night does is make you sit there and take
a moment. And a picture. That's what everyone does, me
included. 'Red sky in the morning' means something's going
to go wrong. So, if you're a sheep farmer, every single morning
should be red.

Make hay while the sun shines

Too right. Don't mess around and take the mickey with your
timings. As soon as that sun's out, get on. Your whole year
could be ruined if you think, 'It should be OK until tomorrow.'
No farmer ever thinks like that.

Under the weather

I can see why this means you're poorly, because the weather
can literally make you sick. So much rain. It's terrible.
What I usually say if I'm ill is that I've got 'mixie' – short
for myxomatosis. The other day I was in the farmyard and I
could tell I was better because my brother turned on the car
headlights and drove towards me, and I jumped out the way.

Take a rain check

A new one on me. I'd have thought it meant going to have a look at the weather and see how much rain fell overnight. Apparently, it means postponing something. Which I hate doing – make hay and so on.

Rain before seven, fine by eleven

This is true. If it's raining overnight and still going at seven, by eleven it'll usually be lovely. And if it isn't, I'm going to start telling people I'm 'taking a raincheck' just to see the puzzled look on their faces.

When the dew is on the grass, rain will never come to pass

Maybe, maybe not, but it's still no good for haymaking. Dew is annoying. You never go baling hay before ten or eleven o'clock because dew's still on the grass and it'll make it too moist.

Cloud nine

This just means that you're happy. I never use it. I think gangsters use it.

When the wind is in the east, 'tis neither good for man nor beast

I've got a terrible sense of direction, so I don't know if that's true. But I hate wind anyway. I don't mind rain or snow, but the wind really annoys me. Rattles me.

It's too cold to snow

That one's true. I say it all the time. People say, 'Brrr, it's going to snow.' I say, 'Shut up, you've been reading too many news articles. It's just freezing, that's all.'

'Dunno, bro, this looks a lot like snow to me.'

Chapter Five

I never thought I understood art, even though I took it as a GCSE. (To this day, I don't really know why I did that. I must have thought it was an easy way to get through school.) Someone would say, 'Look at this amazing painting!' And I'd think, what about it? I don't know what I'm supposed to be looking at. But as I've got older, I've realized it's all about what *you* see in the picture – it doesn't matter what the artist meant when they created it. They scribble a few lines or throw some paint at a canvas, thinking their own thoughts, but to me it means sheep falling off a cliff. It doesn't matter if it's a picture of a ship or a portrait of some posh bloke in old-fashioned get-up, all I see is sheep falling off a cliff. I'm guessing it's because that's what I want to see.

It's … it's so beautiful.

So, I've come to the conclusion that art isn't about what the artist paints, or sculpts, or makes. It's what other people see in it. I said this to somebody the other day and they told me

there's a whole major theory of art based on that idea. As far as I'm concerned, I invented that. Yeah, I may have invented it long after some other people also invented it, but that's true of loads of great ideas. So, I'm claiming it. The Kaleb Theory of Art: art is in the eye of the beholder. And as the beholder, I always just see sheep falling off cliffs.

As regular readers will know, I have a photo of a sheep standing on a cliff, which I only keep because I'm hoping something weird and supernatural might happen to it and I'll get to see the sheep fall off. Other than that, I don't have a lot of paintings on my walls. I like looking at old-fashioned farming paintings: the windmill with the cow in the background and a nice stream running through the picture, that sort of thing. I like art that shows the history of farming. An old man ploughing a field with horses – that tells you how expensive farming has become. When all you had was one horse, you just needed to feed the horse every day. Now you've got to run a tractor – breakdowns, fuel, a man sitting in the cab. Apart from anything else, you don't have money left over to buy paintings at a squillion quid a go.

We're no zoologists, but we're pretty sure the animal on the right isn't a horse.

CONCEPTUAL ART TO THE RESCUE

What I didn't realize before, is that there's more to art than painting pictures. Anything can be art. You just have to point to it and say: 'That's art, that is.' It might still be sh** art, but who's to say for sure? Unless we're talking about an Italian guy called Piero Manzoni, who in 1961 quite literally sealed up his own sh** in tins and sold it at the market price of gold. It doesn't sell at that price anymore, of course. In 2015, one of the tins, which weighs thirty grams, was auctioned for £182,500. At the time, thirty grams of gold would have cost you around 800 quid. This is called conceptual art, and loads of people take the p*ss out of it and say it's just a scam. Including me, to be fair, in the past – but not anymore. I say anyone who can sell his own sh** for 200 times the price of gold is a bloody genius.

*As our American friends would say: sh**canned.*

EXCELLING IN THE FIELD

I've told the story before about how I went on a school trip to London to visit an art museum and never got off the coach. I used to think that meant I didn't like art, but now I realize it was the city that was the problem. It wasn't the art that scared me off, it was the big buildings it was stored in. When it's surrounded by concrete and about 15,000 guards and all kinds of security equipment, I don't want anything to do with it. But when it's in my environment, I love art – it's amazing.

There's got to be an easier way to visit a museum than this.

THE REVIEWS ARE IN

What I've only recently come to understand is, I've been an artist all along. I'm a modern artist who uses farming as my medium. I often get a creative impulse when I'm farming. I'll plough a field in a certain way because I think it'll look cool. Or especially when I'm topping – which is cutting off the top part of a crop to boost its growth – that's really visible, and you can see what you're doing behind you. I sometimes try to draw something. When I go up the field, it folds the grass one way, and when I go down, it folds it the other way. That leaves black-and-white stripes, because of the way the sunlight bounces off them differently. So far, I haven't had any art critics come and appraise my work, but I've had the odd farmer say, 'That looks smart!' I don't try to draw anything in particular, but when you're rolling a field with a flat roller, so you bruise the grass to make it shoot up another stem and double your yield, you also leave lines. The second you mess it up, that's it though, then the whole field's messed up. You've really got to pay attention. But when you get it right, it's awesome. I feel like a true artist when I finish a field. I look back and I go, 'Damn, I'm good at my job.'

'I'm awestruck by the beauty of that field! Such draughtsmanship! Such artistry!'

THE ART OF FARMING

So, I've decided what I need to do is put on an exhibition. They have an Art Week round here and there's a farm near me that's got a lovely wood. In previous years, they've taken all the wood that's fallen down and made it, for instance, into a massive centipede. So, I'm going to show a beautifully ploughed field. I looked to see if artists have an awards ceremony, and turns out they've got loads. The big one's called the Turner Prize. So, why not go for that? I could be the first farmer to win it, at least that I know of. When you're not part of the art establishment or you're not trained as an artist, they call you an 'outsider artist'. But I am trained – I did that GCSE, after all – and I want to be an insider artist because that's where the money is. My artwork, *In a Ploughed Field*, should be worth a million pounds. And the brilliant thing is, I can do it all again next year. Watch out, all you famous artists whose names I don't know. Hang on, I'm going to look some up. Watch out, um … Tracey Emin and Damien Hirst, because I'm coming after you.

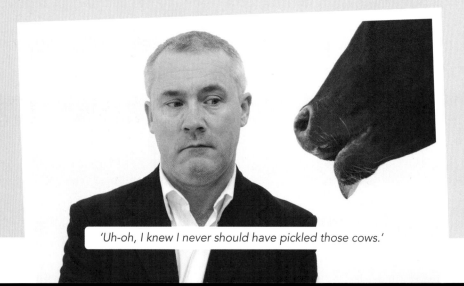

'Uh-oh, I knew I never should have pickled those cows.'

Try to tell me that's not a thing of beauty.

OVERHEAD VIEW

I've often thought of writing a message in a field that people can see from the air, but every time I think of one, I get worried it might offend somebody. I did have one idea, which was to cut some letters in a field spelling out WELCOME TO EDINBURGH, so anyone who was flying over on the way to Luton airport would think they must have got on the wrong plane and start panicking. But then I thought it wouldn't really be fair on the flight crew, and it might even cause an international incident ... So, I gave up on that one. Obviously, the big temptation is to draw a massive penis – I wouldn't be the first. People have been doing that in the countryside since prehistoric times – unless it was somebody more recent doing a prank to make everybody think that ancient Britons liked drawing them as much as we do. (Not that I would ever try to fool people that way myself. Ahem.) I don't know why, it just seems to be a natural impulse to draw them. If you do it on a dirty car window or the pavement, you're a vandal, but if you do one thirty-six feet long on a chalk hillside, you're the representative of a mystical and deep-rooted rural culture.

'Seeing as you ask: it is a club,
and I'm pleased to see you.'

10cc WERE RIGHT

10cc were a band who did a song that goes 'Art for art's sake – money, for God's sake!' There's nothing wrong with putting the two together. As a farmer, I love doing two jobs at once. So, if I know I'm ploughing a field and making art at the same time, that's just perfect for me. You can train crops to grow in a certain way, too. If you grow melons, you can put a cardboard box around them and they grow into a square. Or you could make a pumpkin into a triangle. If topping is a way of making pictures on the ground, then this would be more of a sculpture.

As you can see, I'm already getting to know all the terms. Agri-sculpture, that'll be my thing ... as long as I can sell it after. And it's got to be eaten or it'll go off, and I hate waste. Although, one of the big questions of conceptual art is what art is actually used for. There's an art project inspired by that Piero Manzoni guy, called Black Gold, which sells compost in jars, at the market price of gold, just like his own er ... compost. So, do you put it on your veg patch and get some very expensive courgettes? Or do you keep it as an artwork in the hope that the price will go up? The more I learn about art, the more I love it. I just need to get the right people to love it with me. Ones with loads of money.

FARMING ART CHIC

Now I've decided to be an artist, people might think it'll change my lifestyle and maybe even my personality. But I don't think it's going to make any difference. I think my CV's just going to get longer. I might mix up my style a bit too, though. Grow a moustache and wear one of those weird hankies around my neck, as well as a farmer's cap, just to keep it real. If I start dressing one hundred per cent like an artist, people will just go, 'What a tw*t.' I don't think I'm cut out to look like an artist, though. I think I'm basically designed to look like a farmer. I honestly don't remember the last time I wore something other than a checked shirt. Even my pyjamas have got a check pattern.

I'm not sure if I'll start going to gallery openings and drinking white wine and talking about the exhibits, though. Knowing me, I'd be more interested in discussing how they hung the paintings. There's an art to that, isn't there? You've got to put the nail in the right place, otherwise you'll go through a cable or a water pipe. And you've got to find out where the beam is. And, more importantly, you've got to make it level, and that's incredibly hard. I'd probably end up going around with a spirit level adjusting all the pictures. My hat goes off to the people who hang art, more than to the people who make it. They've probably got some angry artist shouting at them, 'Be careful, that's a million pounds!' 'How is that a million pounds?' 'Well it just is – so hang it gently.'

STATE OF THE ART

All artists have to do something called an artist statement. Basically, you start off saying something like, 'My artistic practice consists of ...', then follow it with two or three paragraphs of words that even you don't understand. This is not a problem that farmers have. A farmer statement would just be a list of stuff you've grown or raised that year. But it does remind me of how I would try to fill the space doing the answers in my English exam. I think my artist statement would be:

'It doesn't matter which way you see art; it's still art. This piece of ploughing is art. There's a skill, an art, to doing it. Turning the soil over in a consistent way is no different to holding a paintbrush and painting the sun red or the grass green. There's also an art to the way I hold my tractor's steering wheel, in the eleven-three position. Normally with my knees, because I'm eating with one hand and on TikTok with the other hand.' Artist statement. Done.

An artist at work.

Chapter Six

Work

IT'S A FARMING THING

In an interview a little while ago I was asked the question, 'What do you do for a hobby?' I said that I like to go to farm sales and buy old stuff that I can restore. 'That sounds like work,' they replied. So, I said that I also like to tinker with tractors and engines in the workshop, trying to fix them up into something I can use in my contracting business. 'That sounds like work as well,' they said. We went back and forth for a good five or six minutes until I had to admit it – every single thing I thought was a hobby, that I do for fun or to relax, turns out to be related to work. Ever since that conversation, I've been trying to find a hobby, but I've just found more work. And, finally, I've realized: I have a hobby. My hobby is work.

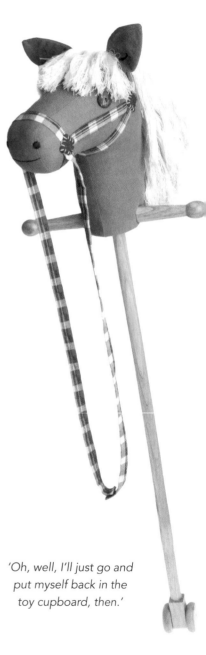

'Oh, well, I'll just go and put myself back in the toy cupboard, then.'

THE KALEB PRINCIPLE

Some people might say that makes me a workaholic, like that's a bad thing. But I don't think so. I think it's one hundred per cent a good thing. The best thing ever, in fact. I've had a motto stuck in my head continuously since I was a kid: 'Dreams don't work unless you do.' Every day, every minute I'm not working, I'm not earning money. That's one of the good things and bad things about running your own business. The more you work, the more you earn (let's ignore the weather for a minute) and when you stop working, you stop earning. It's hard not to feel guilty, thinking you could do more all the time.

'Byeee, Kaleb, we're off!'

I find it hard to switch off at the end of the day. I jump into bed at eleven o'clock and I lie there thinking, tomorrow I've got to go and feed the cows, then I'm going to go and do that bit of topping, and if I finish the topping I can go and jump on that other tractor and move that bit of soil. But it's beneficial for me, because if I keep working, I keep succeeding. And the good thing about farming is that working all the time doesn't keep me from the other aspects of life, like quality family time, which is very important to me. I bring my little boy with me and he loves it! I like to think I'm teaching him new skills, and he enjoys it, being out on the farm or sitting on a tractor. Even my baby girl loves watching the animals, looking at the cows walking up and down. My work life and my family life are the same. And let's face it, breeding labour is cheaper than hiring it.

Dreams don't work unless you do.

JOB VS CAREER

I once heard somebody explain the difference between a job and a career: a job is where you go out, get through the day as quickly as you possibly can, get home and think, right, that's paid the bills. A career is where you work your way along a path or towards a goal that you've set for yourself. I've definitely got a career not a job. I'm very lucky that I found my career so young – at the age of thirteen, in fact. I was fascinated from a very young age by tractor wheels and by animal behaviour, and I wanted to earn some money and help my family. But it wasn't really about money. It was about learning. Every day you learn something different. Farming did for me what school couldn't: it made me want to learn.

So, I knew what I wanted to do and I knew what I wanted to be, and I went looking for any kind of farming work I could get. It's incredibly hard for a non-farming person to get into farming. I would just offer to help and the farmer would say, 'OK, but I'm not going to pay you.' I'd keep coming back and, because I made myself useful, they'd soon start paying me. Farmers are always fair like that. I know how fortunate I am that I didn't lose any time.

'I just haven't worked out what I want to do with my life yet.'

GRANDAD KNOWS BEST

I've always said farming saved me as a person. Without it, I think I'd have been someone else completely – someone I wouldn't like very much if I met him. My mum and dad split up when I was twelve years old, and it was a tricky divorce. I felt like I was in the middle of it. I stepped up at a very young age. Maybe I grew up too quickly. I'm twenty-five years old but I've got two kids and a forty-year-old mind. But if the farming bug didn't catch hold of me the

Kaleb (right), his brother and his grandpa. If this picture doesn't make you smile, either you have no soul or you hate Manchester United more than is good for you.

way it did, I think I'd probably have been a complete a***hole. I was lucky to have a great role model: my grandpa, my mum's dad, Michael Smith. I looked up to Gramp like I've never looked up to anyone else. He passed away in 2021, but I think everything I do – even now – is to please and impress him. He worked incredibly hard. He was a plasterer, and he was also in the fire service, and he ran charities, and he'd help on a farm during the Christmas period. He never stopped. He was always there to provide and make sure none of his family had to worry. So, when I went out farming, for me to see his face and hear him say, 'Well done, Kaleb!', that meant more to me than anything.

MY BRILLIANT (AND CHICKEN-POWERED) CAREER

It all started just as I reached my teens. For my thirteenth birthday, my mum bought me three chickens. I soon realized I could make some money out of selling the eggs. I started selling the eggs locally around Chipping Norton. I worked out that I could sell an egg for 25p and my profits were around 10p an egg. Within two months, I had about 450 chickens. I remember my mum saying, 'I didn't expect this, Kaleb.'

EARLY MISSTEPS

After selling lots of eggs and saving some money, at the age of fourteen – f*** knows why I did this, but I always say, you learn from your mistakes – I bought three sheep. Everyone knows my opinion on sheep. Well, this was where it all began. It made sense to me because I figured, while I was on my egg round on a Saturday, I could take some fresh lamb with me and offer it to my customers. And it sold well. Then, at the age of fifteen, I bought my first tractor.

In fairness, we don't think **anyone** expected this.

The thing is, I'm not from a farming background. My dad, as I mentioned in a previous chapter, is a carpenter and my mum's a dog groomer. It's incredibly hard for someone with a non-farming background to get into farming, and from my thirteenth birthday that was all I ever wanted to do. So, I thought the next best thing to me or my mum or dad having a farm would be for me to have a tractor, and to be a contractor and work on other people's farms. I could still do the thing that I loved the most, only I'd hire myself out. I was so excited to buy my first tractor. It cost me five-and-a-half thousand pounds that I'd saved up. And I swear I ended up paying almost that in repair bills every month. That thing was a complete lemon. It's as if somebody at the factory said, let's make one that breaks down all the time, just to mix things up a bit.

Lad, dog and rusty tractor for hire.

EGGUCATION, EGGUCATION, EGGUCATION

Bear in mind that, while I was doing all this – setting up my own business and running the egg round – I was working full time on a dairy farm as well. I don't like admitting this, but I did drop out of school a little bit. And by a little bit, I mean a lot. And by a lot, I mean almost completely. During Year 9. OK, and Year 8. Yeah, and some of Year 7. I did go back to school in Year 10, as part of a negotiation, because the school kept fining my mum. I said to mum, 'Just relax, I'll pay the fine. I'm earning loads of money. I'll budget for it.' So, I kept paying the fines. But the school got onto us, and brought us in and said, 'Look, we can't have this, come on.' And I said, 'I know we can't. You're costing me fifty pound a week. That's 200 pound a month, 2,400 pound a year.' They said, 'Kaleb, you were never that good at maths before.' I told them, 'I'm learning plenty on the farm and by running my own business.' Anyway, we agreed that I would come back for Year 10 and Year 11 and do my GCSEs. I went back in Year 10 and chose my subjects, but I worked out that everything I was going to do in Year 10 would be recapped again in Year 11 and I would sit an exam on it. In the end, I always went in on a Friday, because I'd worked out that my best egg customer base was my teachers, and I could earn more money selling the eggs over six hours in school than I could in six hours on the farm. So, I'd go in and deliver seventeen dozen eggs to my teachers. Nobody can ever say school didn't teach me anything.

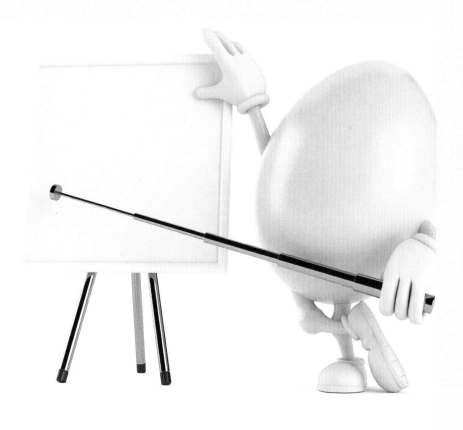

'And now, class, if you'll turn to page 37 ...'

SMART MARKETING

In Year 11 I went back to school properly and finished my GCSEs because I wanted to go to agricultural college. I went straight into Level 2 agriculture, and also did my Level 3, at Moreton Morrell College, which is nearby. At this point, I was sixteen or seventeen, and I was really pushing my contracting business. As my work slowly picked up, the way I marketed myself wasn't to go to the farmers. At that age, if I'd gone to knock on a farmer's door and said, 'Look, I've got this knackered old tractor, how about I come and do your field?', they'd have told me, 'I'm not being funny, Kaleb, but I can get a bigger, newer tractor that actually works to come and do it.' So, I decided my best bet to build up a customer base was horsey people.

Horsey people always put their horse in a lovely stable with fresh bedding. Lots of lovely hay, a lovely water trough, lovely food, some lovely carrots – lovely everything, basically. But in the daytime, when they go to the paddock, the horses are poor grazers. They always leave stinging nettles and dock leaves. They overgraze one patch but leave the rest. They always sh** in one massive area and never eat there (who knew horses were familiar with old Mafia proverbs?).

NOW THERE'S LOVELY

So, I would go to the horsey people and say, 'You keep your horses in that lovely stable all night long, but then you leave them to graze all day in that horrible paddock.' And they'd say, 'Ooh, yes, Kaleb, how do we fix that?' I'd say, 'Well, I come in, I top it for you and I chain harrow it.' And that is how my contracting business got started. I was still working full time on a dairy farm as well. When I was eighteen, I left that dairy farm and started working for an arable farmer who was contract-farming Jeremy's farm. So, that was my day job, and in the evenings and nights I put some LEDs on my tractor and did my own contract work. I was still milking cows in the morning. I could either sleep or earn money, and I wanted to earn money and expand everything as much as possible. The only reason I've stopped is because of my family, so I can have a play session and some breakfast with them in the morning. Now I'm twenty-five years old, farming about 3,500 acres, we've got seven tractors and a team of six people. Dreams don't work unless you do – so I do, and they do too.

There's a saying that goes: 'Find a job you love doing and you'll never work a day in your life.' I've done that and I can tell you, it's complete bollocks. Everything I do is bloody hard work. There are lots of jobs in the world that need doing and nobody really wants to do them – and plenty of them are in farming. Some people reckon that if you become a farmer you've got to

'I thought I was busy, but this Kaleb guy is something else.'

shovel muck all day. Well, they're not lying, I'll tell ya! But while I'm shovelling, in the back of my head I'm thinking, all of this is going to go on my field and produce some really good corn. Or I look at the animals that produced it and I think, I know they're going to be well looked after because I've cleaned up properly – I've done a thorough job here. And let's face it, it keeps me fit. So, my good fortune is, I never resent it. Never once have I woken up in the morning and thought, 'Oh no, I've got to get up and go muck out them cows.' If I did, I'd know I was in the wrong job instantly.

'I see this as less a chore, more a vocation.'

WHERE THERE'S MUCK THERE'S BRASS

What makes a smelly, difficult job worthwhile is understanding the value of it. I know the worth of what I'm shovelling. Cow muck is like liquid gold – OK, semi-solid gold. I don't have to go down a mine like they do in some places, I just go into the cow yard. I've made a business out of moo poo. I've spread a lot of muck in my time. We cover about fifty farms at the moment – and I mean literally cover them. I get paid to go and spread poo. There's not a lot of people outside politics or the media who can say that – and mine delivers much more positive results.

'Mwahahahahahahahaha! Mine! All mine!'

Chapter Seven

Machines

I love machines. When you like farming as much as I do, you're always going to love machines. Every day I'm grateful I was born in an era where there's a machine for almost every aspect of agriculture, and they're all incredibly cool. To me, at least. I know a lot of kids think cars or bikes or scooters or whatever else are the business, but that's only because most of them have never been on a combine harvester.

What a beauty.

TRACTORS

Anything with wheels and an engine and hydraulic pumps and hydraulic rams, I will definitely adore it, but there's never any question what's right at the top of the list. The greatest feat of engineering, the pinnacle of human endeavour, the monarch of all that is mechanical: the tractor. When I think back to what got me into farming, at a young age, it was tractor wheels. Other kids might be singing about how the wheels on the bus go round and round, but it was the tractor wheels that stole my heart.

'You are so Deere to me ...'

To this very day I'm still obsessed with watching tractor wheels driving through mud. I don't know what it is. I can just sit there and watch it all day. I admit, it's weird – and even weirder to see it written down – I feel like I'm confessing to some kind of fetish! Remember that MP who got kicked out of Parliament for looking at porn while claiming he was searching for tractors? Well, if that had been me, I actually would have been searching for tractors.

Corrrr…

MORE TRACTORS

So, tractors were the first thing that got me hooked – my gateway into farming. When you're thirteen you don't get the opportunity to ride a tractor straight away, because you're not old enough and you've not got the knowledge to drive the big kit. But I quickly fell in love with working with animals – which I was old enough to do – and that meant I got to watch the tractors. Some kids want to be astronauts. Some kids want to be firefighters. I wanted to be a tractor operator. Maybe it needs a different name to make it sound as awesome as it really is – 'tractor pilot', for instance. And here I am, ten plus years later, living my dream. I've got seven tractors now – literally one for each day of the week. I've got a tractor from every major brand in my fleet (... I love saying that word: my *fleet* of tractors). I've even got an old Lamborghini, which I used when I was just starting out with my business and which I don't tell anyone about because I slagged off Jeremy's Lamborghini. I mean, what kind of idiot would buy a Lamborghini tractor just because the name sounds flash? Ahem. But at least I've got the excuse that I was only fifteen years old.

Come on, though, it's got that supercool bull logo and everything.

KALEB'S TRACTOR TOP TRUMPS

I got my first ever tractor when I was fifteen years old. The bragging rights were amazing – just fifteen with my own tractor. And it was back in the day when I was courting too, and it was wicked as far as that went. But the tractor itself – a Case International – was rubbish. It kept breaking down. So, I got the Lamborghini and that was much better. Then I bought two tractors from CLAAS, which didn't live up to their name. They were costing me £5,000 a month to keep them working. That's enough to buy two new tractors a year. Those tractors almost finished my business, and then a New Holland tractor pretty much saved it. So, here's my handy Top Trumps-style guide to tractors I have owned.

Case International 240 AFS

Courting potential: 9
Bragging rights: 10
Reliability: 2
Value for money: 5

Lamborghini Premium 1060

Courting potential:	10
Bragging rights:	10
Reliability:	8
Value for money:	7

CLAAS

No CLAAS pictures allowed in this book.

Courting potential:	2
Bragging rights:	2
Reliability:	1
Value for money:	1

Ford 4000

Courting potential:	10
Bragging rights:	10
Reliability:	8
Value for money:	10*

*It's now considered a classic, so worth more every year.

New Holland T7.210

Courting potential:	6
Bragging rights:	10
Reliability:	8
Value for money:	10

AIN'T NO PARTY
LIKE A TRACTOR PARTY

Normally, a farmer will think, well, I love John Deere, I'm gonna buy John Deere for the rest of my life. Or Massey Ferguson, or New Holland. And farmers are very competitive about it. Young farmers' parties are just a tractor fest, with everybody arguing about how their favourite brand is the best. But because I've got one of everything, I can agree with all of them: 'Oh, yeah, I've got one of those.' It's brilliant. I'm like the tractor king. I never have to buy myself a drink.

'Hooray, Kaleb has arrived – time to start the party!'

DIGGERS

I love diggers but, I can never get my levels right. I guess I don't enjoy it as much as working on a tractor, and when you don't enjoy something you lose your concentration. I can sit on a tractor for eighteen hours and I won't get bored, but three hours on a digger and I'm fed up with turning around in circles. My focus is gone, my mind is wandering, and the massive pile of soil that I'm trying to level out looks great in my head but is bumpy as hell in reality.

Some people have what I call 'the digger's wrist', but I just don't. Which is a shame because it would be pretty useful. If you have it, you can just jump in a digger and drive it smoothly. The joystick in a digger moves in a different way from anything else you might

Never let Jeremy near the digger.

be used to. But if you have the digger's wrist, every time you pull it, it'll go the right way. For example, Jeremy does not have the digger's wrist, and he'll push that joystick the wrong way every single time he gets in a digger – before working out the right way.

LOADERS

The same goes for loaders. How did anyone get anything done before diggers and loaders? The loader is basically the farmer's mobility scooter. Every farm I go to, there's usually the dad or the grandad in the loader, and if you ever need anything

done, he'll do it with that. They use it for moving everything, and if they don't have to get off it, they won't. 'Can you move this chair, please?' 'Yeah, I'm on my way.' 'No, I don't need the loader, I just need you here moving it with your hands.'

I don't blame them, though. Anyone who has been working on a farm for more than a couple of decades has shifted enough weight in their time without any mechanical assistance to be cut a bit of slack. I was working with a farmer the other day, and he's an old boy, and he said something about 'fifteen hundredweight'. I was like, 'What is that?' 'Oh, we used to lift fifteen hundredweight by hand.' 'Yeah, but how much is that?' It turns out that one hundredweight is a fifty-kilo bag. They used to pick them up and carry them just like that. You wouldn't dream of lifting that these days. Everybody says, 'Health and safety, your back ...' To which I reply, 'Too bloody right.' Here's to health and safety, and my back, and machines that lift stuff for me. I'm glad I was born in 1998.

'Trust me, this is even less fun that it looks.'

MOWERS

People do their mowing with a lawnmower and for some reason they like to boast about it: 'My deck's 53 inches!' I love joining in on those conversations, 'Oh, really? Mine's three metres.' I don't have to do my lawn with a lawnmower. That's a machine for amateurs. I cut my lawn with my tractor and my topper. It's just a normal-size garden so the topper's almost as big as the lawn is. I like going to gardening clubs – my Nan loves gardening and all her friends love talking about it. I've never been to a gardening club with her, but she's the one who gave me the idea. So, now I go to these clubs and imagine her listening in: 'I've bought this new mower, it's got a 52-inch deck, it mulches, I can sit on it and turn it around.' 'Oh, nice. I've got a three-metre one that does exactly the same job.' I do have to admit, this is one situation where my passion for machinery doesn't necessarily bring out the best in me.

I fought the lawn, and the lawn lost.

MOTORBIKES

There aren't many machines I have a grudge against but I do hate motorbikes. I think it's because my uncle was killed in an accident riding a motorbike when he was only sixteen, so I know how dangerous they can be. When you're a farmer it's hazardous enough dealing with all the machinery involved in your work. (I know I said I'm glad to live in a high-tech age, and I stand by that, but back in ancient Egypt nobody ever fell into a threshing sieve and had their arm torn off.)

At least there's no chance that this will chew your arm off.

My point is, I'm not about to get on a machine that makes life more dangerous than it needs to be. Especially when every single time I jump on, I always, always fall off the bloody thing.

I just cannot physically drive on two wheels. Gravity and balance are teaming up to tell me something, and I'm listening. OK, you can pop a wheelie and all that. Wicked. But is it really that fun? OK, yes, probably – but still. Biker songs make it all sound very romantic. They call it an 'iron horse', and that's great, but a horse has four legs. A car has four wheels. A tractor has four wheels. A quad bike has four wheels – it's right there in the name: 'quad', from the Latin, meaning 'one in each corner so you can't fall over sideways without some serious effort.' (I may have taken some small liberties with the translation there, but that's pretty much the essence of it.) I do have a quad bike, for farming stuff, and it's all right so far as that goes, but even that's dodgy enough, so I'm not about to basically cut it in half and ride the result. Four wheels good, two wheels bad.

'You wanna come over here and say that, boy?'

CARS

Cars are definitely my favourite non-farming machine. Obviously, I work with Jeremy, who's a car nut, but I have my own car – which I like – and he has his own cars, which he likes. The main differences are: one, he likes green and I like white and black (my car is white with a black bonnet and a black roof and – SPOILER ALERT – a black spoiler). And two, his usually cost the same as a four-bedroom house in Mayfair, wherever that is, and mine set me back a few thousand quid. It was the same when I was at school. Everybody was like, 'I want a Ferrari' or 'I want a Bugatti.' And I was like, 'Well,

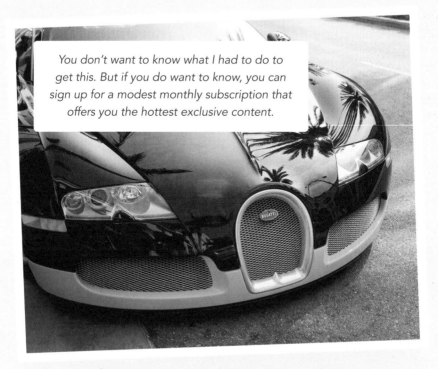

You don't want to know what I had to do to get this. But if you do want to know, you can sign up for a modest monthly subscription that offers you the hottest exclusive content.

I want a Nissan 350Z.' And three years ago, I got it. I can drive around in my Nissan right now. Meanwhile, what happened with the lad who wanted a Bugatti Veyron? They're two million quid, mate. You'd better go and start up an OnlyFans or something.

My personal classics are from the 2000s. Anything between 2000 and 2009 is what I want to collect now. If you go to a classic car show, everybody else will be looking at all the exhibits, but I'll be in the car park going, 'Ooh, I fancy that, and that – and that.' The teenagers are buying those cars right now because they're cheap. And some of them have got pretty big engines. I'm not being horrible, just realistic, when I say that putting teenagers together with big engines, well, they're going to write off a lot of those cars – and increase the value of mine. So, when I get to sixty years old and everybody else now thinks these cars are classics too, and they're worth a little bit of money, with any luck I'll still have mine in the shed.

Just look at those sweet, sweet rides.

UTILITY VEHICLES

Pickup trucks are just awesome. A lot of farmers rave about the old Land Rover Defender, the 110 as it was first called. Admittedly, they look wicked and I'm sure they were the top British agriculture brand back in the day – but have you ever sat in one? It's awful, like climbing inside a tank on wheels, which is basically what it is. Amazing for its time – but I get instant claustrophobia and the moment it starts moving all my fillings get rattled out of my teeth. When I'm checking the sheep, I want to do it in luxury because it takes the edge off having to do anything sheep related. You know that they're going to have found some new way to kill themselves, so at least what you're driving should give you a little bit of joy.

With the pickup, you've got the comfy cab in the front, room for kids in the back seat – because obviously you've bought a

'Right, how fast can this thing get us out of here without shaking us to death?'

OK, but what about a double-cab
Range Rover pickup though?

double cab model – and you've got all the space in the world
in the back to carry whatever you like. You don't want to put
a dead sheep in the boot of something like a Range Rover, do
you? I know they brought in the Range Rover to be a kind of
luxury version of the Land Rover, but then what happened?
Drug dealers bought them. It's the British vehicle of choice if
you're selling cocaine. If you see someone driving a pickup, you
can trust them. You know that if you follow them, you're going
to end up at a good pub. But if you follow somebody driving a
Range Rover, you'll probably end up down a back alley having
your wheels stolen. If you follow a Land Rover Defender, you'll
end up in a farmyard, obviously, but if you follow a Land Rover
Discovery, especially a black one, you'll find yourself on a
film set or in the grounds of a posh private school. So, it's the
pickup for me, every time. See you in the pub!

Chapter Eight

Parenting

Parenting has been much on my mind lately. By which I mean, it's been impossible to think about anything else. When you've got one toddler and one infant, you can forget about sleep. Or sanity. Or clean clothes. Or watching the telly for more than seventy-three-and-a-half seconds at a time. You'll even forget what your own name is. And exactly what on God's green earth possessed you to do this to yourself.

I spend all my time at the moment thinking, 'I can't wait 'till they're eighteen', or 'I can't wait 'till they're thirty', or 'I can't wait 'till they're fifty.' Then you sit there and tell yourself, 'Hang on, what's that about enjoying them while they're young?!' No matter how desperately I may long for oblivion's sweet kiss to come to me and offer me merciful release, I remind myself, 'For God's sake, Kaleb, you've got to be the best dad ever. #blessed.'

Life before having children.

Life after having children.

ANIMAL PARENTS AND WHAT I LEARNED FROM THEM

It's always amazed me, to this day, to look at birthing animals and at animal parenting, as well. I suppose it's the reason we're all here – any kind of animal, us included. Not to eat as much as possible and get really fat, although I'm happy to give it a go, but to breed and reproduce. But nobody ever teaches you the things you need to know about having kids. There are no lessons in schools about how to be a good parent. Everything I've learnt, I've learnt off a cow, a sheep – OK, so that one's mainly what not to do – or a pig. And I've got to say, so far, it seems to be paying off.

CHICKENS

Silkies and pekin bantams were the first animals that taught me about parenting. Silkies are a breed of chicken with amazing fluffy plumage, the kind that I had when I was a teenager (the chickens, that is – my own amazing fluffy plumage is a more recent development) and they are brilliant mums. You have to make sure a chicken gets enough calcium in its diet because otherwise it'll dissolve its own bones to produce eggshells. The mother chicken has to sit on the eggs for twenty-one days straight, making sure there's just the right amount of humidity and the perfect temperature, only leaving the nest once a day to do its business and have a little bit of

Not only remarkable parents, but also dead ringers for your favourite Seventies' rock band.

food and drink, before returning. When the chicks hatch, they raise them with such care: finding them food, calling them over, teaching them to scratch and forage.

SHEEP (INEVITABLY)

Because I had my own chickens, I could see first hand what an amazing job they do bringing up their young. Then came sheep. That was my second experience of watching parenthood close up, and it certainly showed me the other side. I only had fifteen sheep – Hebrideans, a hardy Shetland breed.

Dunno about parenting tips, but it's clear where Kaleb gets his hairstyle ideas from.

Sheep have their own personalities, just like humans. We've got one sheep called Stumpy and I can't sell her because my other half wouldn't let me, which I suppose has been good training for those times I'm tempted to ask if we can put the kids up for sale. But Stumpy isn't like the rest of the sheep. She is a fantastic mum. She is also the exception. The other sheep, like the one called Edith … oh my God. When she was lambing, she just wanted to kill you. You could say she was being protective – the sheep equivalent of, 'If you come near my kids I swear I'll do time,' but as soon as she'd lambed, she'd be like, 'I don't want that baby anymore. Take it away, not interested.' We had to foster the thing every time. Then there was Hazel who gave me a headbutt and a fat lip when she was poorly and I was trying to drive her and her three babies up to the barn. Admittedly, I probably shouldn't have been transporting her in the back of a VW Polo, but that's what you do when you're seventeen years old and you can't afford insurance on a truck. She had a lovely comfy seat, she had a seat belt, she had air con. I even made a little ramp so she could walk in and out – but there's no pleasing some sheep.

Sheer luxury, if ever we saw it.

COWS

A two-year-old cow that's never had a baby before – a heifer – will give birth and know exactly what to do: get up, turn around, lick the calf until it's dry, stand there quietly to let the calf find its feet, then teach it to suckle. How does that mother know all these things? It's not like they get taught how to do it. I'm sure they're not listening to their own mother and taking notes. Or going to cow school.

Then again, maybe they are.

MY PARENTS

I can imagine my mum and dad – my mum especially – hearing me say that I learnt everything I know about parenting from animals and blowing a fuse. Fair enough, I guess, because of course I learnt loads from my parents, too. My mum was always very loving and I was a real mummy's boy. Both mum and dad worked incredibly hard – they still do, all the time – so I think I get my work ethic from them, but the most important thing they showed me is that you can work and still

have fun with your kids. They bought four-and-a-half acres in Heythrop, a village near Chipping Norton. They invested in it, building a stable block and all sorts of other things – but, when I look back, I think, was it really just about being an investment? I think it was even more about being good parents and getting us kids outside, doing stuff as a family.

So now, when I farm on a daily basis – when I feed cows and bed them down, for instance – I always take my little boy with me. He loves it, and so do I. And I can thank my parents for showing me that by their example.

YOUNG FATHERS

I always wanted to be a younger dad. And I've always said I think I grew up too early because of my mum and dad getting divorced – I'm only in my mid-twenties, but in my head I'm more like thirty-five or forty. Some people might say I look it, too, but never mind that. By the time I really am thirty-five, my little boy will be twelve, and I'll still be young enough to be out there working my arse off, trying to get to where I want to be. And if they want to, my little boy or little girl can join me on that road, and I'll still be able to enjoy it.

HOLDING THE BABY

I'm very lucky that I found the right person at a very young age. I've been with Taya, my other half, for nearly eight years now and we're getting married before too long. Most importantly, we both wanted to be young parents. And it's amazing how that natural instinct I saw in the animals I look after just kicks in for us humans, too. Before you have kids, if somebody hands you a baby, you've got no idea what to do. You panic. Which way up does it go? Am I holding it so hard I'm going to leave dents in it, or so gently that I'm going to drop it? But when it's your own baby, somehow straight away you know what to do. I couldn't explain it to anyone else though, but I'd just tell them to trust themselves – you'll know what to do when you need to.

Awwwwww.

LABOUR PAINS

My respect for women was high in the first place. But when I saw Taya give birth to Oscar, our firstborn, my respect doubled. And then doubled again two weeks later when she said, 'Oh, I'd love another one.' All I could think was, I've seen what happened to your body, I've seen you screaming, I've taken a punch from you ... and you want another? Evolution must have given us some kind of amnesia that affects people who've had a baby, otherwise they'd never, ever have another. Taya said:

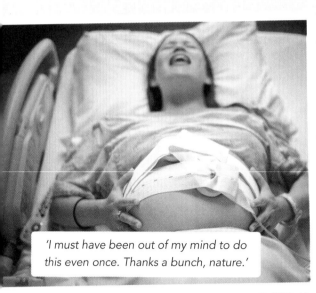

'I must have been out of my mind to do this even once. Thanks a bunch, nature.'

'I've just forgotten the pain. I can't remember what it feels like.' The only reason the human race didn't die out long ago is that women are tough as hell. I've been farming since I was a kid, I've been kicked, bitten, butted, stamped on, trampled, you name it, by animals, and had every kind of injury from tools and machinery, but you can put it all together and it doesn't even come close to what women go through in childbirth. Then they go right ahead and do it again! Heroes, I tell you. I am in awe, frankly.

MILESTONE MOMENTS

People talk about what it's like when your kids start crawling or walking, and it's boring if you've not had kids – but when it is your own child, it really is like magic. Especially as you don't know where they're going to end up. In about five seconds flat, they go from being exactly where you put them to being anywhere within a five-mile radius. You'll look up at a shelf that you can't get to without a stepladder and there's your toddler, happy as Larry. How did they get up there? Levitation? A jet pack? But for all of that, there's nothing quite like hearing your child's first words. In the case of my kids, it's remarkable they even got a word in edgeways. I went on tour recently and I was told I had to rest my voice beforehand to prepare for it – which was when I realized just how much I talk. It's a lot. The good thing about being told to rest my voice was that I had to learn to shut up and listen more. Having a conversation with my own son is incredible, especially when, just for once, I'm babbling less than he is.

'Nobody looking? All clear? Right, ceiling cupboard here I come!'

THE TOUGH TIMES

There are moments when you want to cry. I think any parent would be lying if they say there haven't been times where you sit there and think, I can't do this anymore. I remember my little boy had a temperature of over forty degrees and was lying there limp, and I was terrified: 'What on earth do I do?' Even when it's not that serious, when it's runny noses and sore throats, you'll go for weeks without having slept properly and it all seems too much. You feel so helpless. When you take the longer view, you can console yourself with the idea that they're building their immune systems, but when you're in the moment it can feel brutal. In some ways, I think we've got off a bit lightly – I'm convinced my kids don't get as ill as other kids, and I once read that children who grow up in the countryside have stronger immune systems than those who grow up in cities. It makes sense because my kids are outside on the farm, they eat a bit of dirt, go and lick a gate every now and then ... all in the name of a good cause! I'd just like to add, by the way: whoever invented Calpol is an absolute genius. If they haven't been given a Nobel Prize, then I want to know why.

But, no matter how difficult it gets, I wouldn't change it for the world. Parenthood is by far and away the best thing I've ever done.

'Mm-m, lovely textured mouthfeel, earthy flavours balanced with a sprinkling of mulch and a crisp sandy coating. Now to lick a gate for afters.'

Chapter Nine

Fashion

Before television came along, it never really occurred to me to think about fashion, but I'd like to say that, since I started being on TV, I've become more fashion conscious. Unfortunately, that would be a lie. They say you should dress to impress – but dress to impress who? I'd already got a girlfriend and she knows how I dress (which makes it quite amazing that I'd already got a girlfriend, but I'm not going to complain

about my luck). So, basically, I'd be dressing to impress cows. And trust me, cows don't give a toss what you look like in the morning when you go and feed them. There's plenty of things they do – quite literally – give a sh** about, as anybody who does my job can confirm. In spades. Specifically, the spades you use to clear up the cowpats. But fashion isn't one of them.

'If we can make an effort, Kaleb, we don't see why you can't too.'

Fast-forward to 2019 and we started filming the first series of *Clarkson's Farm*, and one of the production team said to me, 'Do you always wear that checked shirt?' And the answer is 'yes'. I never wear polo shirts. I never wear a plain black T-shirt. Some younger farmers do, but that's a youthful folly. When they turn twenty-five, they'll be straight into a checked shirt and that'll be that for the rest of their lives. I never went through the black T-shirt phase in the first place. I've been wearing checked shirts since I was twelve or thirteen – although the rumours that it's been exactly the same shirt all along are false and malicious. I've got another one as well. Same with my gilet, jeans and my workboots. 'So, what do you on special occasions?' Well, if I'm going out for a nice meal, I just put on a cleaner set of all that stuff.

My dream walk-in wardrobe.

YOUNG KALEB'S STREET STYLE

When I was a kid, I just wore whatever my mum put me in that morning, then I went out to cause havoc. When Chipping Norton had roadworks and we were coming home from buying sweets, I thought, what better way to end the day than to put all the 'ROAD CLOSED' signs on the main junctions? Havoc,

I tell you. It was all good until the poor council got it in the neck. Oops! But the point is, I didn't require any particular outfit to do that stuff – unless you count joggers, the traditional garb of the underage troublemaker.

Waiting for the casting call from Farmers Weekly *– the farming version of* Vogue.

THESE BOOTS ARE MADE
FOR KER-CHING!

My other half Taya's got a real thing for fashion. It's developed over the last five or six years, and her wardrobe is enough for both of us put together. She's obsessed – and I cannot stress this enough, *obsessed* – with Fairfax and Favor. It's a British company that deals in real leather, and she's got so many boots that I can't even begin to explain the extent of it. It's as if we've got a whole herd of cows in boot form. It's lovely but it's very expensive. I'm the opposite. I've only got those few items of

clothing I mentioned because who needs more? You wear it, you wash it, you put it back on. You don't need three entire wardrobes full, let alone so many boots that it looks like you've started a breeding programme. I sometimes think I hear them walking about at night, whispering to each other in some weird shoe language, and plotting to take over the sideboards and the kitchen cupboards. If Taya had all the money she's spent on boots, she'd probably go right out and spend it on boots again, so I think it's best just to let that one lie. Also, I'm told it's a business started by two young blokes when they were in their teens, so I have to admit, I wish I'd thought of it first.

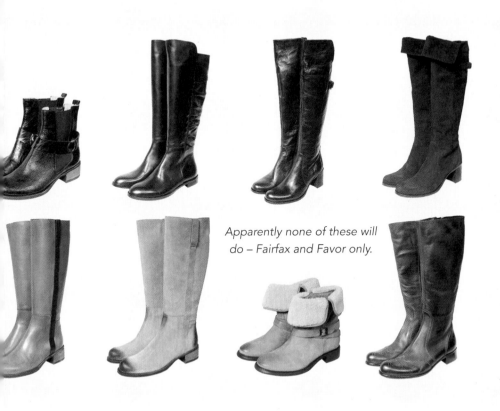

Apparently none of these will do – Fairfax and Favor only.

A FIELD GUIDE TO SPOTTING A FARMER

One useful thing about fashion is that you can always tell a hobby farmer from an actual farmer by the way they dress. First, a farmer will have a good pair of wellies, an expensive pair because they're going to spend all day in them and cheap wellies don't end up being cheap – they're a classic false economy. But a real farmer's wellies won't be a designer brand either. If you see a guy in Hunter or Barbour wellies, straight away you know he's somebody trying to look like a farmer. He's got four acres and four sheep.

Second, real farmers are never clean. They're not scared of getting dirty and they don't care what they look like. Hobby farmers are always really, really clean. They wear expensive jackets and they like taking pictures of themselves in them, because they want to document their life and they've got friends in the city that they want to show off to. A real farmer will never take pictures. They simply haven't got time. OK, it's true that I get my picture taken all the time, and I even get filmed – I wouldn't be writing this if I didn't – but somebody else does the photography and filming. I just get on with my job and let them get on with theirs. If I stopped to take a selfie every time I saw a baby animal, I wouldn't be wearing any kind of fashion at all, even my kind, because I wouldn't be able to afford it.

... and even the sheep have got wellies.

FEMALE FARMING CHIC

Women in farming basically wear the same things as men – the checked shirts, the jeans and the workboots. The only difference is the style of the jeans – my jeans are blue, and so are most male farmers', but a lot of the women prefer black jeans. They usually have a headband on too – I'm jealous of that. Nothing to stop me wearing one, I suppose – maybe I should plan my next hairstyle around it, David Beckham style. The trouble is, I like things a bit loose and those hairbands are very tight and constricting. It might cut off the circulation to my head and then where would I be? It's the same with tight skinny jeans. My voice goes up an octave just thinking about it.

Men over 35 in skinny jeans.

Farming wardrobe, his 'n' hers.

THE WINDS OF CHANGE

If I could change anything about farming fashion, I think I'd have summer wear and winter wear. I'd like jeans for the summer that let a little bit more of a breeze through them. There's a couple of new pairs of shorts that have become popular, and one of them in particular have really taken the need for breeze seriously. Every young farmer that I know wears these shorts, to the point that I think, 'This brand must be making a killing.' Again, I'm annoyed I didn't think of it first. Even worse, I can't physically wear them because they're so short that things fall out that really, really, one hundred per cent need to stay inside. I don't want to expose myself to that kind of trouble. I'd love to have a bit more of a breeze cooling things down – but there are limits, and I'm a strong believer in sticking to them.

'Kaleb, what are the chances of sourcing some Diddly Squat branded shorts within a 16-mile radius?'

FARMING FASHION REVIVAL

There are some older farming fashion lewks I'd like to see come back. I'm not talking about the Middle Ages or anything – although farmers were pretty cool then, too, if the Internet is anything to go by. I think the hobby farmer vs real farmer thing was happening even then.

'Out of the way, nobleman, peasant coming through.'

The classic farming look – wax jacket on point.

No, I'm thinking about the 2000s. Waxed jackets are one thing I'm really sorry to see fall out of fashion. When I was growing up, everybody had a wax jacket. The people that still wear them now have the same ones they did then. But the people who weren't around then aren't buying them anymore. Farmers have got updated clothing with higher-tech fabrics that don't require so much maintenance, and I can understand that. When you're outside in all weathers, having a jacket that is rainproof and windproof is a big deal, but I do love the old waxed jackets. The reason they're going out of fashion, let's face it, is you've got to re-wax them. I want a coat that I can put on and know I can trust to be waterproof forever. Still, those old jackets are beautiful. I love the look of them. It's one of the few times I actually start to care more about what something looks and feels like than what it does.

Nowadays though, waxed jackets are more popular as a fashion item with hobby farmers – the 'green welly' brigade – and also, for a while, with football hooligans. Weird combination. If you saw someone in a classic Barbour, you couldn't be sure if they were going to hire you to top their horse pasture as they didn't really have a clue what they were doing, or try to kick your head in. Mind you, I wouldn't fancy the hoolies' chances against an equal number of young farmers. Or even old farmers for that matter. Especially old farmers. Those guys are tough as old boots – which is just what you'll get, right in the breadbasket, if you start anything with them.

KALEB COUTURE

I'm starting my own clothing brand. Given everything I've said about not caring about fashion, you might think that is a strange thing to do, but that's the point – even people who don't care about fashion should have decent things to wear. So, I'm just taking the stuff I like to wear day-to-day and doing it the way I'd like it to be. I want to use really good materials, British wool and so on, and have it all be proper quality. I'll have beanies, checked shirts and three-quarter zip jumpers (like a cross between a jumper and a cardigan). I'll also have a pair of work trousers that look like jeans but are better supplied with pockets. I always need somewhere to put my pocket knife, a bit of twine, my phone, my wallet – and I don't want them all in the same place. I want enough pockets to hold each thing securely. Talking of pockets, you've got to have pockets you can trust. I'm going to bring some boxers out, too. It's all very well worrying about what's on the outside, but as people always say 'It's what's on the inside that counts.' Another thing I'm going to have: shorts that your bits don't fall out of. Because, if it's what's on the inside that counts, then you probably want to keep it there. The main reason I'm doing it is for other farmers – clothes made by a farmer, for farmers, because we know what we need – but if hobby farmers want to look like real farmers, they can always buy some, too. They'll feel better about themselves, trust me.

Not gonna lie, that is some pretty sweet merch right there.

Chapter Ten

Exercise

Exercise and I have had an on–off relationship for years. Status: it's complicated. I have mixed feelings about exercise, and I think exercise is pretty ambivalent about me, too. 'Hey, exercise, shall we get together over the weekend?' 'Sure, let's – but I'm not gonna lie, I will hurt you.' I've never been into anything kinky but try telling that to exercise. 'Had enough, you bad boy?' 'Yes. Yes, I have.' 'Oh no you haven't.'

KALEB THE GYM BUNNY

You might not think it to look at me now, but I used to go to the gym all the time. Back when I was dairy farming, I would go three times a week to a local independent school, one of those posh ones that have their own gym, swimming pool and sauna that you can join. So now I can tell people, with no word of a lie, that I went to private school – and unlike Jeremy, I never got expelled.

Working with these beasts is a decent workout in itself.

THE TRACTOR WORKOUT

But since I moved into doing more arable farming and my company got really busy, I just haven't had the time, and I've put on a bit of weight, sitting on the tractor more and more. Obviously, I think tractors are the best things ever, but I have to acknowledge, they don't do much for your figure.

Who needs a protein shake when you can have a pork pie?

BODY SHAPE ISSUES

For a long time, body image has been a big worry for women. Everything they look at or read tells them there is one definition of beautiful, which I think is really unfair and wrong. More recently, it's become a major issue for a lot of men, too, and that's not good either. I've got a few friends who love going to the gym and, of course, people should do whatever they like and enjoy – just don't take it too far and think that it is the most important thing about yourself.

I do sometimes see guys developing anxiety about how they look. Mental health is an important issue to me, one that I talk about a lot – I've even done a charity single for the Royal Agricultural Benevolent Institution, called 'I Can't Stand Sheep'. What a lot of people don't realize is that the rate of suicide among farmers is really high. In 2021, male farm workers were three times more likely to take their own

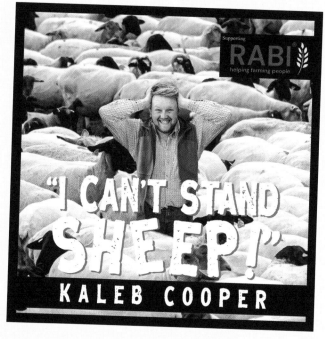

lives than the national male average. All credit to Jeremy for speaking out on this issue as he has done.

There are a lot of things that contribute to that statistic. Loneliness is a big one, of course. I think for a lot of young men in general, not just in farming, body image is a factor, too. They see things on TV and think, 'I must look like this', 'I must look like that.' I'm lucky in that it doesn't worry me. I'm just a normal person: I wouldn't say I'm exactly unfit but I've definitely got a bigger body build now. If I want a six-pack, I'll go and buy half-a-dozen cans of cider. Or sell half-a-dozen eggs – it never fails. I don't care about that, as long as I'm healthy enough. I'm basically a bear, a brown bear. In the winter, I put on weight and hibernate. In fact, I've been thinking of nominating myself for that Fat Bear Week competition they have in the USA, but I'd have to send in a picture of myself catching and eating raw salmon, and that feels like a bit of a stretch when I spend most of my days in a field outside Chipping Norton.

RUNNING

If a cow gets out, I can still run faster than said cow and catch it. That's the standard I set myself. People sometimes ask me, if I'm so good at running, why don't I do more of it? There's a simple reason – running is unproductive. I have enough work to do that demands my energy – I'm not going to start running for the sake of it. It's the same with weights. 'Do u even lift, bro?' Yeah, I do that all the time, I have to, so why would I want to pick something up and put it down fifteen times in a row, then put it back where it came from? I've got hay bales and sacks of feed to move. If you gave me a set of dumbbells and told me to lift them, I'd say what any farmer would say: 'Wait here a minute, I'll go and get my loader.' So, the only times I run are when I'm chasing something, or something's chasing me – as long as I can run faster than either of those 'somethings', I'm doing fine.

.BEWARE OF.
THE BULL

His time over 100 metres is better than yours. Come to that, it's better than Usain Bolt's.

NUTRITION

So, what I don't do in the gym, I make up for on the farm. There's a saying that exercise is medicine and that people who do physical activity tend to have better physical and mental health than people who don't – and I believe all that's true. But I also know that the food we eat has a massive impact too, so I have someone who helps guide me on food. She explains what's good for me to eat and what's not, and what I should be taking in on a daily basis for my workload. At the very beginning, she told me to think of my body like a car. If you put twenty miles' worth of fuel in and drive twenty miles, you'll use up all the fuel. But if you put thirty miles' worth in, that extra ten miles' worth will just sit there. So now I'm more careful, especially in those fat-bear winter months when I'm less active, otherwise I'd just eat, eat, eat, eat, eat. I watch what's going into the tank. Maybe more bears should have a nutrition coach. Or maybe, if you really want to get fit, you should hire a bear as a personal trainer. I can guarantee you'll be breaking every track-and-field record in the book on a daily basis.

'Dude! Today you will improve your personal best. Or be eaten. Your call.'

EVOLUTION VS ABUNDANCE

Food's taken for granted these days – we can pretty much buy the food we want, whenever we want, even if that means flying it in from the other side of the world. I think we should do more seasonal eating and maybe not eat as much. It's hard for us to motivate ourselves to exercise or to eat less because that's how we've evolved: to eat as much as is available, because food used to be scarce and finding it was hard and dangerous work; and to do as little as we possibly could, so as to conserve energy and not burn off what few calories we had found.

Now, we can get food in a shop or delivered to us any time we like – as long as we can pay for it, of course – and, as

The paleo diet was not all it's cracked up to be.

a result, we waste huge amounts of it. We throw away more than nine million tonnes of food a year in this country alone, and around the world over one-third of all the food produced goes to waste. As one of the people who produces it, I find that shocking, especially when so many people in this country are in food poverty, and so many people around the world are hungry or even starving. I like to make money as much as the next guy, but I don't believe in producing food just for it to be swept off a plate and into the closest bin. I believe in producing food to feed people, and I think all of us could be a lot more careful about what we eat and what we waste. I say it over and over, but you can't repeat it too often: in the end, it doesn't matter what phone you have, what car you drive, what kind of house you live in; it matters what's coming across your plate for breakfast, lunch and dinner. Even if the last two things are really the same thing – but let's not open that can of worms again.

Guess what's for dinner (at lunchtime)?

THE KALEB DIET AND THE KALEB WORKOUT

Given that the farm is basically my gym, I've got my entrepreneurial brain thinking: why not team up with my nutrition coach, so she provides people with an eating plan for a healthy diet, week-by-week, and I supply them with the food they need for each week's plan, straight off the farm? I can really see that working. Unfortunately, because my entrepreneurial brain never knows when to stop, it went on to another idea: why not treat the farm as a literal gym? People can pay a membership fee to come and work here. I get free labour – better than free, they actually pay me – and they get the benefit of a highly effective fitness regimen. I'll even give them personal coaching. I know I don't exactly look like I'm qualified for the role but you can carry a bit of timber and be quite fit underneath it at the same time. I can still go out and do a full day's work on the farm and not complain about it, which I bet a lot of really hench types couldn't do.

'I started the Kaleb Workout only this morning, and I weighed 275 pounds then and couldn't see my own feet.'

'I've never felt more empowered!'

LIFTING BALES

This would be the first thing I'd teach you in the Kaleb Workout. Always bend at the knees. In the summer months, you've got to handle bales of hay, and it's really hot, so you sweat like a pig. (Which reminds me, I must add wrestling a sweaty pig to the workout routine. No point doing anything with sheep, unless it's picking up a dead sheep because, as you know by now, sheep have only one ambition in life and that is to die. But I can always market that as a deadlift.) Anyway, I digress: you've got to pick up the bales and put them into hard-to-reach places while getting a rash …
quite honestly, I don't think I'd want to pay for that. But then I remember that there are people who pay good money to enter insanely difficult and uncomfortable cross-country races, or go to boot camp, and I can offer an experience every bit as gruelling and miserable. I mean, ahem, challenging. 'Challenge yourself with Kaleb's Hay Bale Workout.' You'll definitely learn something about yourself. Mainly, that you're the kind of person who'll hand over a load of cash to perform a task that most people wouldn't do at gunpoint.

COW CHASING

It's hard to schedule this one because you never know when the cows will get out – but when they do, it'll be handy to call up my fleet of people to go and chase them and bring them back. 'Right now! Yes, I know it's midnight, but this is a crucial part of your training. The cows are running all around Chipping Norton, let's go and get them back in.' The uncertainty means they'll have to pay me a retainer every month, of course, just in case it happens. That's only fair. Plus, it maintains the element of surprise essential to stopping any workout plan becoming routine and plateauing.

Come on, girls! A dozen gullible fitness freaks are counting on us!

POTATO HARVESTING

If I grow a field of potatoes and I want to harvest them, I can use a machine to do that, which costs me money. But if a hundred people come to my fitness course and everyone has to bring a spade or a shovel, then they can just dig their way across the field. With potatoes, I usually do them in rows of three metres across. But at one metre per person, I could do a hundred metres in one field. Farmers boast all the time when they get in the pub, 'I have a three-metre cultivator.' 'Oh, mine's six metres.' When I go to the pub next time, I'll be able to say, 'Yeah, my potato combine? A hundred metres long, mate. Bet you can't beat that.' And at the end of it, the clients have to buy the potatoes back because it's part of their health plan. If they're not fit by the end of that, they're never going to be. OK, your back may be screwed, but your body fat and muscle tone and overall physique? Like a Greek god, mate.

In hindsight, I should have insisted they used a different Greek god.

Chapter Eleven

New Age

Usually when I hear the term 'New Age', I think of all the robots and other high-tech stuff that's coming in. There's been a big fuss about artificial intelligence, AI, and it seems to have everybody terrified. Not me, though. First off, if there was a bit more real intelligence around, maybe we wouldn't need the artificial kind. I say, bring it on. I'd rather have artificial intelligence around me than authentic stupidity. Second, everybody's worried it's going to take their jobs – everyone except farmers. I'd like to see a computerized brain muck out

'Oh dear, I'd love to help but, as you can see, these tomatoes require my full attention, indefinitely.'

a cowshed. No, I mean it. I'd actually *like* to see a computerized brain muck out a cowshed. When can it start?

Anyway, it turns out 'New Age' means something quite different. It's all about mystical ideas and practices. Astrology, spiritualism, druids, that sort of thing. Now, anyone who knows me will probably remember that I had a traumatic experience with some druids, or at least people who looked like druids, or who looked like they thought they ought to be druids, when I was young. I saw something no kid should see

one night at some ancient stones nearby. Those were some dirty old druids is what I'm getting at, and I don't mean they hadn't wiped their feet.

I had another traumatic experience on a farm I worked on when I was fifteen. Lots of hippie earth-loving types would come up there, to be at one with nature – which I can get on board with. They had a hot tub they'd made out of a boat, quite ingenious actually. I jumped in and I was sitting at the pointy end, happily minding my own business, when a lady jumped

There's a reason why for most of his teenage years Kaleb basically looked like this.

in after me. She had the biggest boobs I've ever encountered in my life and she wasn't wearing anything on top, because she was a woman of nature and didn't believe in clothes. They were floating towards me, her boobs, the whole time, and she kept holding them back as we were talking, because they were pinning me in the corner. I know some people might have thought this was great, but I wasn't one of them. I was trying to generate waves under the water to push them away because I was scared of getting punctured by a nipple.

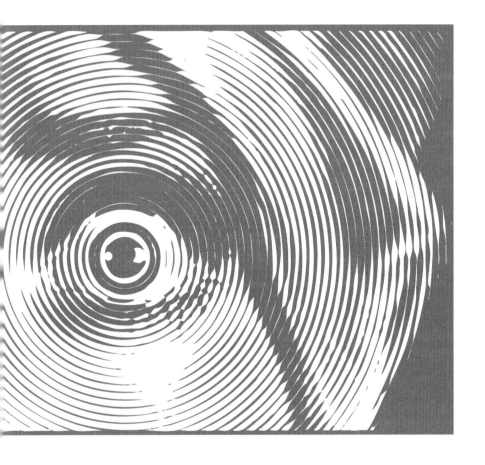

THE SECRET OF THE STONES

For a long time, I never went near that stone circle, the one where I'd seen the druids up to all sorts, if I could help it, but more recently I got interested in the history of it. I was cultivating the field next to it and we had a person with a metal detector – a detectorist, I think they're called – come in before me to go and check I wasn't going to hit anything you might want to dig up and preserve. His buzzer went off, so he dug down and found a bit of Roman jewellery with a pearl in it, so he kept digging and found a skull. It was time to call in the archaeologists. They reckoned it was a fifteen-year-old girl who'd had a head injury or a brain tumour and she'd been buried with jewellery all around her. She'd been laid to rest aligned with the stones, so when the sun rose and set, the light would pass over her body, thanks to the way the rays and shadows are guided by the stones.

At first, I just wanted to get on and plough the field, and I was annoyed because they stopped me. But then I started to pay attention and I found it fascinating. I like history, so I got really into watching and listening to what they were all saying. I like the fact that no one really knows what happened, but everyone's got a different opinion on it. There's something quite touching about the thought that every day, for hundreds and hundreds of years, the light has been passing over that girl's resting place.

I'm not sure myself if it genuinely is a magical place. If I say it isn't, I'm going to upset quite a few people, and if I say it is, I'm going to upset even more people. What I can tell you is, if you stand in the middle of the stone circle, and you put a marker on one of the stones and count all the way around, then use the same marker and do it again, you always get a different number. Always. Even if you get someone to stand on a stone, the same thing happens. So, you can make of that what you will.

'Ten. I count ten of them. Do you get a different number lying down?'

HOMEOPATHY

You might think that New Age ideas are more popular with visiting city folk and hobby farmers than they are with country people, but that's not actually the case as far as I can tell. When I was growing up, there were a lot of people around who were into homeopathy, which as I understand it is taking an ingredient, watering it down until there's almost nothing left at all, then using it as medicine. My boss's parents practised it. So did some of my mum's friends. Given how much they practised, they ought to have been really good at it by then. At the farm where I used to work, they used homeopathic methods to treat their cows, putting drops of something that reminded me of Rescue Remedy in the water – and they didn't have many ill cows, I must admit. [*We would like to point out that Rescue Remedy is a trademarked preparation of botanical essences, and in no way connected with homeopathy – legal ed.*] Now, maybe they just took excellent care of their cows generally. Or maybe the drops helped? I don't know if I believe in it, but I don't dismiss it either. I'm not homeopathophobic. Sometimes, my mum's friend would give me a tiny bottle of potion to take to help me feel relaxed, if I felt nervous before an exam, say. I think it helped. Both me and the cows – the herd and myself – were ready for the exam. I reckon the cows definitely would have done better on it, though.

ASTROLOGY

Loads of people read their horoscopes every day, in the newspaper or online somewhere. It's really a mainstream thing now, but it's never been of interest to me because there's no connection to farming. I wouldn't avoid operating heavy machinery because Mercury is in retrograde, whatever that means. Apparently, it's bad news and things are more likely to go wrong – but I'm not about to start writing things in the Accident Book like 'Sliced half my finger off because of an unfavourable planetary conjunction.' Homeopathy got my

attention because of the idea of using it to treat cows. But I'd never get a horoscope for a cow. I already know what's in a cow's future. It usually involves me having to spend more time than either the cow or I would like in the cow's back end. The ideal amount of time for both of us would be zero, but life doesn't work like that and no alignment of the planets or the constellations is going to change that. That's also why I'm not interested in aromatherapy, because when you spend that much time around the back end of a cow, the less you have to smell, the better. It can definitely have an impact on your state of mind, but I'm not sure it's an entirely positive one.

'I see Kaleb's working on the thresher – let's ruin his day, shall we?'

ACUPUNCTURE

I've never had acupuncture but I'm interested in trying it – I've heard good things. I still have my doubts, though. It's one thing getting pricked with a needle when you're getting a vaccination or something useful like that, but getting stuck with needles purely for the sake of it? I'm not sure about that at all. I spend loads of time giving cows and sheep injections, and I know what I'm doing it for. Whereas with acupuncture it's all a bit vague. A load of needles in me … for what, exactly? This is someone *literally* stabbing you in the back, and what's more, you pay them for the privilege, then say, 'Oh, yeah, thanks that's really helping.' I hear that some people do it in order to stop them smoking, which luckily isn't a problem for me, but I suppose if every time you sparked up a cig somebody came along and jabbed a needle in your shoulder blades you'd be motivated to give up pretty sharpish.

Why not cut out the middleman and target the problem at its source, we say.

PSYCHIC POWERS

I don't know if I believe in psychic powers. A psychic will say, 'Did someone pass away recently?' 'Yeah.' Then they try to narrow it down. 'Was it one of your family?' 'No.' 'Oh wait, yes … yes … I'm getting a message … was it a friend?' The tactic seems to be, keep it broad and ask lots of questions. I sometimes get supposedly psychic messages on my Instagram. 'I'm getting drawn to you. Your spirit is drawing me in. Can we sit down and I'll read your palm?' And I'm thinking, I wouldn't, love, I know where it's been. And you definitely don't want to try reading my mind because it'll just be me running over sheep all day. There is one exception, though, where I totally believe in psychic powers. Your mum, and my mum, and everybody's mum, totally possesses them. From when you're small, the first thing you learn is that your mum can read your mind. Mums just know. I don't know how they know, but they know. They look at you out of nowhere and say, 'Don't even think about it.' And you don't so much as question what they mean because you were always thinking about doing something you best not do. You just think, 'She got me. Again.' Chalk up another one to Mystic Mum.

Mums know
EVERYTHING.

TAROT

Definitely not for me, this one. I don't like scary or spooky stuff as a rule, and this is nothing but. A pack of cards full of figures like Death and The Hanged Man? I find ordinary playing cards weird enough. I mean, have you ever stopped and looked at the King of Clubs? That is one hard-arsed geezer. The guy's got a double-edged short sword and a stare like you just dropped a hay bale on his favourite dog. And don't get me started on the Ace of Spades – a card that's even got its own theme song.

So, you can keep the Burning Flageolet and the Nine of Coffee Spoons and all the other creepy business you get in a tarot deck. I'm going to stay away from them bad boys, thank you. 'You're not supposed to take it literally' is what they always say, and thank God for that, because otherwise Death is going to hang me upside down from a fiery tower while the Devil crouches in wait for my soul. Although at least the Devil looks like a goat. If he looked like a sheep, now that would be truly terrifying.

We've been to worse stag dos, to be honest.

SHIATSU

Not, as I initially thought, something my mum would need to take a lot of time and trouble over at her dog grooming business, this is a kind of Japanese massage where they use their knuckles, elbows and feet to reduce tension and reenergize the body. I've got to say, that sounds bloody awesome. I love a massage. Deep tissue massages are my favourite, but this looks extreme and extreme is what I need. My work gives me knots in my muscles you couldn't get out with a toffee hammer. They can have a crack at me any time they like. I hope they want a challenge, is all I can say.

Wrong.

YOGA

No. Absolutely not. I can't sit still for longer than two seconds, let alone hold a pose and stretch my arse out. My body's not shaped for it. Just like my mind isn't made for meditation. I don't think I want to spend too much time alone with my mind – the thought scares me. My mind is a dangerous place. It may well do me some good, though. I can't switch off and that's why I have trouble sleeping at night. I'm always thinking, what can I do to grow the business, to get to my dream of owning my own farm quicker? What if I set up another business? The entrepreneurial spirit in me doesn't let me get any rest. And I'm worried that yoga and meditation, instead of calming that spirit down, would give me more opportunity to think about business and unleash even more entrepreneurism. For instance, right now I'm thinking, what if I diversified to meet New Age demand? I could start selling eggs from psychic chickens, or cut runic symbols into somebody's field with my topper?

Wait! I've got it! Crystal healing for foot-and-mouth disease!

Chapter Twelve

Nostalgia

There's a book called *The Go-Between,* which I like the idea of, even though I haven't read it. First, because it's set in the English countryside, and I always like anything that celebrates the nature we have in this country. But second – this is the reason I've heard of it – because somebody told me a line from it once: 'The past is a foreign country: they do things differently there.' That's an awesome quote because it's so true. Especially for me, never having been to a foreign country yet. But I can still imagine what it's like to go abroad just by thinking about the past and how different so many things are now. Farming life in a lot of ways has got more difficult, and the area where I live has become really popular with hobby farmers – although that's turned out quite well for me in one way, as everybody knows. But in another way, it's made things much harder, because it means it'll be more difficult to get a farm of my own eventually.

The book was written by a guy called LP Hartley and if you say his name to older people – you know, anyone over thirty – they always start babbling on about an advert for the Yellow Pages, whatever they are. Anyway, that's a different Hartley, who apparently was into fly fishing.

'My name? Hartley, JR Hartley – licensed to fish.'

BYE BYE BIRDIE

Even the fairly recent past feels like a foreign country to me now because, even though I'm only twenty-five, I've got a lot of responsibilities and I'm always working. When I look back to my teens, I think of the times when I didn't have things to worry about, apart from the amount of trouble I could get into – not too much trouble, just enough to

'I suppose you think this is funny, do you?'

make it fun and be remembered for years afterwards. I was never the ringleader, never really a troublemaker, but I did knock about with people who were. When I was in college, we used to go to the local golf course and drive golf balls down the driving range – but with a slight twist. We'd get these things called 'crow bangers'. They're like a little dynamite stick to scare birds off and they make a sound like a gunshot. We used to strap them to the golf balls, light them and then hit them down the run. They'd be going straight – then the crow banger would go off and they'd suddenly veer in a different direction and cause a little bit of chaos. Again, just the right amount. Until they go through a window, and then you'd need to have it on your toes pronto. Still, what do you expect if you live next to a golf course?

I never really took part in the very naughty stuff as a teen. I was just there, enjoying the results. I'm not like a 'do it' person. I'll just watch and laugh. There were often times when I felt like saying, 'Hang on, this is too much.' But did I? Of course not. You *think* it, but you don't say it. Unless we were about to get caught – I was very good at piping up then. But we almost never got told off or punished. I had a knack for assessing the situation and a good sense of timing, and when I think about the people I hung out with, I bloody needed it.

'Right, lads, call it a sixth sense, but something tells me it's probably best we leave it there.'

GOING HOG WILD

If you farm from a young age, like I did, your playground is the farm. My old boss knew that and occasionally when we'd finish work, he'd say, 'All right, go jump on the quad bike, put the helmet on and fly round the farm as much as you like – let off some steam.' Every farmer, I think, still has that play in him or her. So now, when I'm the one in charge, I'll say, 'Go and get the clay trap out, let's do some clays.' Or 'I wonder what it would be like to ride that pig.' Next thing you know you've jumped on the back of a pig – and got thrown off very quickly, because they're incredibly strong, and you end up in a mound of dirt (and the rest). But it's playtime, still. You're never too old to ride a pig. Too big, maybe; it has to be fair on the pig, otherwise it's animal cruelty. But normally they're twice the size of the largest person there. When you work on a farm you find it's the older generation – the farmers' dads – who do more of the playing and the joking, while the farmers themselves are quite serious. You have to be, to keep things running. But you know that underneath it, they can't wait to get back to the kind of fun they had when they were younger. They're only holding themselves back for the greater good. So, whenever I think back to how much fun I had as a very young farmer, I can always remind myself that, if I'm lucky, I'll get to do it all over again – and really annoy my kids.

*Young Kaleb
(artist's impression).*

*Old Kaleb (also artist's impression
but we think the first artist took
a little more time over theirs).*

THE OLD TUNES

I remember when I first started listening to the radio in the tractor. Normally, the boss would have BBC Radio 2 or Radio 4 on, but I would play Heart or Capital or Radio 1. Now I've moved to Radio 2 myself, which you do as you get older. I may be twenty-five but in my mind I'm more like forty-five. I've gone early.

I didn't ever listen to any country music back in the day, like I do now. I was a massive fan of Akon. And not just him, it was all hip hop for me then. I loved it. I still do now, in truth. When

They see me rollin', they hatin'.

you want to get in proper gangsta mode you just whack that on. Hip hop was the music of my youth. But that all changed in 2016 when I went out there courting. I found rap music wasn't, let's say, conducive to romance. If you turn up rapping and gangsta popping, it doesn't necessarily have the desired effect. They don't write them like they used to, and thank goodness for that. If you listen to the words again now it's very, very inappropriate. That's when I started getting into country music, which in my experience is better suited to courting – and to farming. I mean, the clue's right there in the name.

THE GOLDEN AGE IS OVER

Nostalgia is all about believing that there was once a Golden Age. When I look back at my own childhood, I suppose that was mine. I miss having no responsibilities. When I was eight or nine, going right back, the only thing I ever had to worry about was my fear of balloons, which made birthday parties a bit of an ordeal. I just didn't go. I believe in facing your fears, but not that one – it's too scary, f*** that.

I used to spend every break time at school playing football, as well as six-asides after school. Even primary school itself was pretty awesome. St Mary's Primary School in Chipping Norton was a great place; the other pupils, and the teachers, they were all brilliant. That's another thing that feels like a foreign country now, a place you can never go back to. Soon after I went to secondary, there was no more football after school, no sitting down in front of the telly – unless it was to watch David Attenborough so I could learn about animals. If it was my weekend off, I might play video games. Usually a farming simulator – not like I was obsessed or anything. Or if my mates were playing *Call of Duty* I might jump on it, mainly so I could hang out and talk to all my friends. But mostly I just went to work. And that tied in with my parents splitting up, because that was the thing that made me do it.

If I'm being honest with myself, I miss when mum and dad were together. It was a massive change when they broke up. It felt as if everything we'd all been working up to went away in that instant. When they bought that bit of land

My brother is a cow-hugger like me.

we all worked on as a family, for instance. When my mum and dad were building the stables and putting up fences, my brother and I would get out our toy diggers and dig holes all day. Dig a hole, cart the soil away, come back and dig another one. Great fun. In a way, that's what I've been trying to do ever since – to build something. I felt as if it was left up to me to make it happen. I don't blame anybody. It was nobody's fault. It's just sh** when you're a kid and your family falls apart. You remember everything. Your life divides into before and after.

I'm sure most children of divorced parents feel the same way. One day everything's happy and safe, as if you're in a childhood paradise. And the next day it's all changed. It's strange even now, just thinking about it. I reckon that's why I grew up too quickly. Having my own business at thirteen, trying to help pay the rent. I know a lot of people get nostalgic for their teenage years, because they didn't have a lot of responsibilities and they were having fun – parties, romances, all the stuff that's normally part of teenage life. But my teens weren't like that, for the most part, because I was so involved with building a business and making money for my family. It was as if I went straight from being a little kid to a sensible grown-up overnight, the way people used to before teenage life became a rite of passage. So, for me, the Golden Age is when I was a little kid: mum and dad were still together and everything was easy then.

BANTER R.I.P.

I do miss the old days of banter. The bigger the company gets and the more business it does, the more I've got to be the one at the top who takes things seriously and isn't always up for fun. The one who's saying 'Go go go go go!' when everybody else wants to stop stop stop stop stop and have a laugh. It used to be me playing about, waiting for the boss to say, 'Right, let's get on with some work', and now it's me who has to be the bad guy. Long and short, I'm growing up. I'm the one who's changing, but in a natural way. It's like it says in the Bible – which is something else I haven't read – 'When I was a child, I spake as a child, I understood as a child, I thought as a child: but when I became a man, I put away childish things.' Apart from potato fights, or throwing stones in the river to splash Jeremy while he's building a dam, or riding the occasional pig. Although I'm not sure if the Bible goes into specifics. Bet you didn't think I was going to quote the Bible at you, did you?! Nor did I, to be honest.

THE FUTURE'S BRIGHT, THE FUTURE'S KALEB

Farming changes almost daily. I'd like to go back to a time when the milk price was really good and every dairy farmer was having an enjoyable time – and when the government didn't keep making up schemes that were nonsensical. But I'm not always pining after the 'good old days' – some of the changes have been positive. Ten years ago, nobody ever spoke to farmers. They'd just look you up and down and ignore you. Now, people want to speak to the farmer, be friendly to the farmer. I don't know if it's because then, when it snows, they can get their car towed out of a drift – but I guess that's nothing new. So, I think people are more appreciative of what's on their plate and want to know more about where it came from. I think *Clarkson's Farm* has made a real difference, too. It's got people thinking about farming, seeing it in a new light, and has given viewers a better understanding of the work that goes into keeping up a farm, as well as how difficult it can be to make a decent sum from it.

All that said, while the methods, technology, economics, rules and schemes seem to be constantly shifting and changing – the people remain the same, they don't change. I can guarantee that if you went back to the same college today, the kids there will still be lighting crow bangers on the golf course … although they probably had to source a fake ID to get their hands on them.

So, when people talk about how much they miss the old days, I think what they really mean is they miss who they were in those days. It's not the world that's the problem, it's just you and the passing of time. Better to make the most of the place you're in: the present, with the future open ahead of you. You can't change the past but you can definitely change the future. Make it a Golden Age of your own. So, nostalgia can wait. I've got too much to do.

Nostalgia, shmostalgia.

ACKNOWLEDGEMENTS

A big thank you to each and every one
of the hard-working farmers out there!

Picture Credits

apter One: 11 iStock / gerenme; Chris Terry; 12 iStock nage Source; 13 iStock / Lukas-Shots; 14 Shutterstock Jnberrer; (below) iStock / k_samurkas; 15 iStock / n-balvan; 16 iStock / Griffin24; 17 Julyan Bayes; Chris rry; 18 iStock / Lukas-Shots; 19 Shutterstock / Nattika; iStock / Qwart; 21 Alamy / Marc Hill; 23 iStock / 1Vozd; 24 Author's own; 25 Shutterstock / Enessa rnaeva; 27 Shutterstock / Umomos.

apter Two: 29 iStock / stevecoleImages; Chris Terry; (above) Author's own; (below) Shutterstock / QBR; iStock / Víctor Suárez Naranjo; 32 Author's own; Chris Terry; 34 Shutterstock / Csanad Kiss; 35 iStock lamedeeso; 36 iStock / NNehring; 37 Shutterstock / vid Beauchamp; 38 Shutterstock / Audrey Snider-l; 39 iStock / GlobalIP; 40 iStock / xalanx; 41 iStock / teKaras; 42 iStock / GlobalIP; 43 Author's own.

apter Three: 45 Chris Terry; 46, 47 Author's own; iStock / MrKornFlakes; 49 iStock / master1305; Chris rry; 50 Chris Terry; 51 iStock / Gannet77; 52 iStock / drey Bukreev; 53 iStock / monkeybusinessimages; Author's own; 57 (above) Shutterstock / Dana.S; low) Ellis O'Brien 59 (above) Alamy / Pez Photography; low) iStock / Bulgac.

apter Four: 61 iStock / Romolo Tavani; iStock / Ljupco; ris Terry; 62 iStock / Eugene Sergeev; 63 iStock \ntagain; 64 Alamy / Chronicle; 65 iStock / Yoela; Daisy Waterman; 67 Ellis O'Brien / Prime Video; Shutterstock / Ritu Manoj Jethani; 69 Julyan Bayes; Shutterstock / Sergey Novikov; 71 Chris Terry; 72 iStock ubaphoto; 73 Julyan Bayes; 74 iStock / panaramka; Author's own.

apter Five: 79 iStock / Ljupco; iStock / Fajar Sunny; ris Terry; 80 iStock / MediaProduction; 81 The eam Edvard Munch, 1893, The National Museum Art, Architecture and Design, Oslo, Norway: Alamy ncamerastock; 82 iStock / clu; 83 Shutterstock / ve Smith 1965; 84 Shutterstock / Andrey Burmakin; Dreamstime / Andrew Roland; 86 Shutterstock / Sodel dyslav; 87 iStock / Sjo; 89 Shutterstock / Roberto La sa; 90 Shutterstock / Carmela Fasano; 93 Chris Terry.

apter Six: 95 iStock / DNY59; Chris Terry; iStock / EdnaM; 97 iStock / alexsl; 99 Chris Terry;) Shutterstock / Stokkete; 101 Author's own; 3 Chris Terry; 104 Dreamstime / Adrianadh; Author's own; 107 Shutterstock / Talaj; 110 (above) eamstime / Alphaspirit; (below) Shutterstock / Thomas cher; 111 Shutterstock / SCP Photography.

Chapter Seven: 113 iStock / PhonlamaiPhoto; iStock / VR_Studio; iStock / Igor Borisenko; Chris Terry; 114 iStock / ArtistGNDphotography; 115, 116 Author's own; 117 Dreamstime / Tycson1; 119 Author's own; 120 Shutterstock / William A. Morgan; 121 iStock / shaunl; 122 Dreamstime / Matylda Laurence; 123 Shutterstock / Ljupco Smokovski; 124 iStock / Tofotografle; 125 iStock / Pavel964; 127 Shutterstock / Uncleroo; 128 iStock / LPETTET; 129 iStock / Tolga_TEZCAN; 130 Shutterstock / Ian Redding; 131 Julyan Bayes.

Chapter Eight: 133 iStock / monkeybusinessimages; Chris Terry; 135 Shutterstock / suparna graphic; 136–7 iStock / Iluzishan; 139 Shutterstock / cynoclub; (below) Dreamstime / David Elliott; 140 Alamy / Mr Standfast; 141 Shutterstock / Kashper; 142, 143, 144–5 Author's own; 146 Shutterstock / christinarosepix; 147 iStock / Asian Alphan; 149 iStock / jjshaw14.

Chapter Nine: 151 iStock/ feedough; Chris Terry; 152–3 Dreamstime / Sandchia82; 154 Shutterstock / Kazuno William Empson; 155 Ellis O'Brien / Prime Video; 156–7 iStock / Carol_Ann; 159 Alamy / MediaWorldImages; 160 iStock / master1305; 161 iStock / Simon Skafar; 163 Alamy / Lily Alice; 164 Alamy / North Wind Picture Archives; 165 iStock / John F Scott; 167 Ellis O'Brien.

Chapter Ten: 169 IStock / michaeljung; Chris Terry; 171 Chris Terry; 172 iStock / Mark Gillow; 173 Author's own; 175 Shutterstock / dleeming69; 177 Shutterstock AI; 179 Shutterstock / Denis---S; 180 Shutterstock / Stephen Mcsweeny; 181 Shutterstock / Kurbatova Vera; 182 iStock / rogermexico; 184 iStock / Gerard Koudenburg; 185 Shutterstock / tose.

Chapter Eleven: 187 iStock / sandsun; Chris Terry; 188–9 Shutterstock / terra.incognita; 190–1 iStock / George Peters; 193 Shutterstock / Algol; 195 Chris Terry; 196–7 Shutterstock / Tanya Antusenok; 198 Dreamstime / jochenschneider; 199 iStock / pldjoe; 200-1 Shutterstock / Fer Gregory; 202 Shutterstock /Jagoda; 203 iStock / Karina Uvarova.

Chapter Twelve: 205 iStock / duncan1890; Chris Terry; 207 Shutterstock / Glenn Copus / Evening Standard; 208 Shutterstock /D. Cunningham; 209 Alamy / AJ Pics; 211 (above) iStock / pictore; (below) iStock / OLHA POTSIEVA; 212–3 Paul Nicholls / Royal Agricultural University; 215 iStock / Jens Rother; 216 Author's own; 220 Dreamstime / Adonis1969; 221 Chris Terry; 224 Chris Terry.